_1993_

ALGE BRIDGE ™

Concept

Based

Instructional

Assessment

Individual *Algebridge*™ Units

Fractions in Expressions and Equations

Meaning of Negative Numbers

Pattern Recognition and Proportional Reasoning

Constructing Numerical Equations

Attacking Word Problems Successfully

Concept of Variable

Concept of Equality and Inequality

Operations on Equations and Inequalities

Educational Testing Service

The College Board

ALGE**B**|**R**|**I**|**D**|**G**|**E**™

Concept

Based

Instructional

Assessment

Janson Publications, Inc. Providence, Rhode Island

Printed in the United States of America

9 8 7 6 5 4 3 2 1   3 4 5 6 7 8 9

Design Malcolm Grear Designers

Typesetting Rosenlaui Publishing Services

Illustration Verbatim, Inc.

Contents

Contents

Preface

Algebridge™ is a set of supplementary curriculum materials that, as the name implies, is aimed at helping students make the transition from arithmetic to algebra. It helps teachers *assess* and *instruct* algebraic thinking within the context of arithmetic.

- *Algebridge* focuses on concepts from arithmetic that are needed in the study of algebra but that are not ordinarily stressed in arithmetic and algebraic instruction.

- It emphasizes development of concepts and abilities rather than memorization of algorithms.

- It seeks to diagnose and correct errors in *thinking*, not errors in computation.

Students who understand and can work with the concepts presented in each *Algebridge* unit will have the prerequisite skills they need to succeed in algebra. Thus, *Algebridge* not only helps to enhance the instruction of those students who intend to take algebra but also increases the pool of students who are mathematically ready to take algebra.

The task force that developed *Algebridge* consisted of classroom teachers, teacher educators, mathematics education researchers, mathematics textbook authors, cognitive researchers, and test developers. Within this diverse team there was strong agreement about what students need to know in order to be successful in algebra. The task force created a list of the concepts and abilities they believed to be crucial precursors to algebraic thinking, grouped these concepts and abilities into themes, and then organized the *Algebridge* units according to the themes. The *Algebridge* materials are designed so that students who understand and can work with the concepts presented in each unit will be well prepared for success in algebra.

Algebridge is an extension of the principles of the College Board's Educational EQuality Project to mathematics teaching. The Educational EQuality Project was launched in 1980 with the twin purposes of better defining key goals in secondary education, both in basic student competencies and in academic subject areas, and of widening the access of secondary school students to higher education. *Algebridge* translates both these purposes—higher quality and greater equality—into curriculum materials and instructional exercises for use in algebra—one of the most critical school subjects for students headed for postsecondary education.

The Educational EQuality Project represents a College Board effort to foster reforms in today's educational environment and to renew its commitment to collaborate with various educational institutions on issues of curriculum, instruction, and the climate of learning. The Project has produced several publications addressed to the teaching community; these range from general statements briefly defining academic competencies in six key disciplines (the Green Book), to more elaborate descriptions of curriculum and modes of teaching within the disciplines (in mathematics, the Raspberry Book), to the present *Algebridge* materials specifically developed for use in the classroom. All developmental materials have resulted from the collaboration of educators from the secondary and collegiate levels and the participation and support of a variety of boards and officials in the field of educational policy.

The first-year introductory algebra course is a logical focus of attention since algebra is at the heart of the mathematics needed for college studies. Unfortunately, it constitutes a major hurdle for many students. It is often presented as a separate, deductive system rather than as a natural outgrowth of arithmetic. Algebra has become the passport needed for entrance into college preparatory programs in many schools. Although there is no denying that the abstract nature of algebra makes it an especially challenging subject to teach and to learn, algebra can be tied more closely to students' previous mathematical experience than is usually done. The transition between arithmetic and algebra needs to be, and can be, made easier for most students who have the potential to go on to college.

A distinctive feature of *Algebridge* is that it draws from the rich body of cognitive research on mathematics learning that has emerged over the last decade. This research has shown how students construct their conception of a given topic from their previous experiences with the topic. In constructing this conception, students often work from incomplete or erroneous information. Consequently, their conceptions are interlaced with erroneous notions. These misconceptions interfere with both learning and problem solving. More importantly, they are deeply rooted and resistant to change. Students have invested time and energy in constructing a conception of a topic, and if that conception is erroneous, it takes at least as much time and energy to dislodge it. Consequently, simply telling students that they have an incorrect conception, even when followed by a correct explanation, is usually insufficient for dislodging the misconception. Dislodging a misconception requires the student's active participation. The *Algebridge* materials are designed to diagnose misconceptions and to help the teacher guide students in constructing correct conceptions about a topic whenever misconceptions are uncovered.

Acknowledgments and Permissions

The Algebridge *materials were developed with the assistance of the following Task Force and Development Committee members.*

□   Member of the Task Force
o   Member of the Development Committee

Algebridge Task Force and Development Committee

□ o **Nancy A. Beck**, Mathematics Resource Teacher, Robinson Middle School, Milwaukee, Wisconsin

o **Nell Cobb**, Mathematics Teacher, Manley High School, Chicago, Illinois

□ **Robert Davis**, Professor of Mathematics, Rutgers University, New Brunswick, New Jersey

o **Lucy Garner**, Mathematics Teacher, Los Angeles Center for Enriched Studies, Los Angeles, California

□ **Carole E. Greenes**, Professor of Mathematics Education, Boston University, Boston, Massachusetts

□ **Charles O. Hall**, Mathematics Teacher, John F. Kennedy High School, Los Angeles, California

□ o **Nicolas Herscovics**, Associate Professor of Mathematics, Concordia University, Montreal, Quebec

o **Carolyn Kieran**, Professor of Mathematics, University of Quebec at Montreal, Montreal, Quebec

□ o **Jeremy Kilpatrick**, Professor of Mathematics Education, University of Georgia, Athens, Georgia

□ o **Jose P. Mestre**, Associate Professor of Physics, Cognitive Processes Research Group, University of Massachusetts, Amherst, Massachusetts

□ o **Dorothy S. Strong**, Director, Bureau of Mathematics, Chicago Public Schools, Chicago, Illinois

o **Gloria Williams**, Adviser, Research and Evaluation Branch, Los Angeles Unified School District, Los Angeles, California

o **Donald Zanotelli**, Chairman, Mathematics Department, Bay View High School, Milwaukee, Wisconsin

Algebridge Consultants
Educational Testing Service

Jeffrey G. Haberstroh, Associate Examiner in Mathematics, College Board Programs

Diane J. Pege, Assistant Examiner in Mathematics, College Board Programs

Paul A. Ramsey, EQuality Instruction/Assessment Project Director

Marlene C. Supernavage, Assistant Examiner in Mathematics, College Board Programs

Beverly R. Whittington, Senior Examiner in Mathematics, College Board Programs

Field-Test Sites

The *Algebridge* Development Team acknowledges school districts in the following cities that field-tested the materials and thanks the participating mathematics teachers for their valuable suggestions.

Boston, Massachusetts; Chicago, Illinois; Los Angeles, California; Milwaukee, Wisconsin; Project Prime, Arizona School Districts; Seaford, Delaware

Staff,
Educational Testing Service

The Team also extends special thanks to the following staff of Educational Testing Service for their contributions to *Algebridge*.

Chancey O. Jones, Elizabeth Carrick, Gerry Fox, Patricia Meier, Susan Gafgen, Cathy Ricatto, Test Production Services, Text Processing

Permissions

The *Algebridge* Development Team acknowledges that the glossary of words, phrases, and symbols used in *Algebridge* is copyrighted by Center for Applied Linguistics, 1118 22nd Street, NW, Washington, DC, 20037, in conjunction with Northern Virginia Community College, Metropolitan State College, and Miami-Dade Community College, and is reprinted by permission. Funding for the glossary was provided by the Fund for the Improvement of Postsecondary Education, U.S. Department of Education.

Notes to the Teacher

Algebridge™ is an instructional assessment program designed to supplement pre-algebra and algebra curricula. The materials can be used in various classroom settings to address the thinking skills needed for problem solving, *within the context of the traditional curricula*. Most important, *Algebridge* is designed to promote students' active involvement in their own learning, building their mathematical confidence and abilities in the process. Also, the explicit focus on concepts that textbooks often assume are already understood helps students identify and overcome their misunderstandings.

There is also flexibility in the function units may serve. Units can be used as an introduction to a textbook chapter, as an insertion within the fabric of a chapter, or as the chapter's final activity. Additionally, the teacher may wish to reinforce the concepts presented in an *Algebridge* unit throughout the school year by including questions like those in the unit on chapter tests or in class exercises.

Each *Algebridge* unit addresses a specific topic and is designed to be used independently of other units. However, it is possible to interweave materials from several units. For example, *Constructing Numerical Equations* and *Attacking Word Problems Successfully* are complementary units which could be worked on at the same time. Additionally, the teacher may also choose to use only portions of a unit at a particular time and to return to the remaining portion at a later time. *Pattern Recognition and Proportional Reasoning*, for example, could easily be used in this manner. Finally, the order in which the units have been arranged in the Comprehensive Teacher Resource Book is only one possible arrangement. Teachers are encouraged to use the materials in any order that seems appropriate for their particular purpose.

The *Algebridge* Units

The concepts that are addressed in each unit can be linked directly to an understanding of the conceptual underpinnings of algebra. What follows is a brief description of each unit.

Fractions in Expressions and Equations

Emphasizes the concepts that are necessary to understand how to work with fractions that occur in algebraic expressions and algebraic equations.

Meaning of Negative Numbers

Helps students to develop concrete models in order to strengthen their conceptions of the meaning of negative numbers.

Pattern Recognition and Proportional Reasoning

Helps students to gain an intuitive sense of the concept of function through recognition and generalization of geometric and numerical patterns. Also develops the concept of ratio and a conceptual understanding of proportional reasoning.

Constructing Numerical Equations

Focuses on the relationships between quantities in a word problem and on constructing a numerical equation that represents these relationships. Uses the technique of representation (figures, tables, and charts) to help students understand the situation presented in a problem.

Attacking Word Problems Successfully

Establishes a foundation for the development of useful problem-solving techniques. Emphasizes interpreting the meaning of each quantity in a problem and relating it to other quantities in the problem with the goal of deciding on a solution procedure, carrying it through, and evaluating the result.

Concept of Variable

Introduces the many roles of a variable in algebra. The development progresses from the concept of variable in a numerical context to the symbolic representation of a variable.

Concept of Equality and Inequality

Develops an understanding of the various roles of the equal sign in an equation. Points out the similarities and differences between equations and inequalities.

Operations on Equations and Inequalities

Uses concrete models such as balance scales to help students understand the effects of performing the same operation on both sides of an equation. The concepts associated with combining like terms and applying commutativity, associativity, and distributivity are also explored.

The *Algebridge* Approach

Algebridge employs a multi-step approach to diagnose and correct student misconceptions. The steps are described below.

1. *Assess* whether a student understands a particular concept by administering the **Instructional Assessment**.
2. *Discuss* the given idea while scoring the assessment. Both student and teacher can see exactly where misunderstanding lies.
3. *Instruct* to clear up conceptual misunderstandings and fill in gaps in knowledge using **practice sheets** and **suggested activities**.
4. *Reassess* using the **Follow-up Assessment** to make sure the concept is fully understood.

Assessment

The instructional assessment associated with each unit assesses student difficulties within specific conceptual themes. *Algebridge* units are *not* intended to be used as testing instruments for measuring student achievement. Rather, they are designed to be diagnostic and thus to help teachers provide better instruction.

Administration of the Instructional Assessment instruments is designed to be flexible. The instruments can be given in one class period or over several class periods, depending on the academic level or personality of the students, the purpose that the unit is serving (introduction, part of a textbook chapter, or final activity), portion of the unit being used, or difficulty of the particular topic. The teacher may find that in some cases the entire assessment can be easily administered in one class period. In other cases, the *Fractions in Expressions and Equations* unit, for example, it may be best to administer the instrument in two or more sessions.

The assessment instruments employ three types of questions: multiple choice, multiple yes-no, and free response. The format of each question is chosen so as to assess most effectively the concept addressed. This varied format should also make the exercises more interesting for the students.

The multiple-choice questions are an efficient means to assess students' understanding of certain concepts. The multiple yes-no questions allow assessment of students' ability to recognize equivalent forms of mathematical expressions and equations. Some of the yes-no questions assess whether a student holds contradicting views about a concept. For example, a student might agree that the product of two numbers is 12 but also might agree that the product is an odd number.

The free-response questions offer valuable insights into students' thinking. The diagnostic value of these questions lies in the fact that they ask students to formulate their own answers rather than choose among answers already formulated. Because the answers to the free-response questions will vary, the annotated version of the Instructional Assessment in the teaching materials provides suggestions on how to interpret answers to these questions as well as to the multiple choice and multiple yes-no questions.

Since the format of the questions is probably unfamiliar to the students, the teacher may want to give the students some sample questions before administering the assessment. Alternatively, or in addition, during the discussion of the assessment, the teacher could question students about their reactions to or difficulties with the formats.

Discussion

Since the assessments are diagnostic, teachers are encouraged to lead a discussion on the questions and the students' responses to them as soon as the class has completed the assessment. *Discussion of the questions is central to the* Algebridge *program* since it is through the discussions that students get a sense of their individual strengths and weaknesses and the teacher gets an overall sense of class performance. (The teacher may want to keep a record by placing a check in the appropriate column in a chart like the following.) Through the discussions, the teacher can encourage the students in communicating mathematically, thus clarifying ideas in their own minds and reaching consensus about the answers.

Question Number	How Many Students Answered Correctly?		
	Most	Some	Few
1. a)			
b)			
c)			
d)			
e)			
2. a)			
b)			
c)			
d)			
e)			

There are several ways to handle the scoring of the assessments. Students may write their answers to the assessment on a separate sheet of paper, in a notebook specifically designated for *Algebridge*, on the optional self-scoring answer sheets, or in their own assessment books. The teacher *should not* score the answer sheets since this will discourage discussion and defeat the purpose of helping students to identify and correct their misconceptions. (Although an annotated version of the Instructional Assessment and an answer key appear in the Teacher Resource Book, some teachers may find it convenient to use one test booklet as an answer key.)

Various error patterns will surface during the discussion of the assessment. These error patterns are discussed in the teaching materials. Additionally, the teaching materials include sections called "Focus of Question," which details what each question assesses, and "Analysis of Question," which explains the difficulties students are likely to have with the questions and errors they are likely to make. Also included is a section called "Guiding Class Discussion" which offers suggestions for helping students construct correct conceptions.

Instruction

After the Instructional Assessment has been used to pinpoint students' misconceptions or misunderstandings, the teacher can refer to the practice sheets and suggested activities in the Follow-up Instruction section. Note that this section is organized by concept according to the list given in the introduction to the particular unit. Once concepts requiring additional instruction are identified, teachers can locate those concepts in the Follow-up Instruction section and proceed with the practice sheets and suggested activities given.

The practice sheets have various formats and purposes: the questions may be very similar to the assessment questions; they may probe further or summarize the concept addressed by the assessment question; they may present the concept from a new perspective through a novel problem situation or an application of the concept. Not all students will need to do all practice sheets. The teacher should use the diagnostic information obtained from the assessment and choose those practice sheets and activities which address the particular concepts that were difficult for the students.

The time necessary to implement the instructional materials associated with each unit may take from several days to several weeks, depending on the students' ability to grasp the concepts in the unit, the extent of the students' misconceptions, and the purpose that the unit is serving.

Reassessment

Depending on the function which the *Algebridge* unit has served, the teacher may choose not to administer the Follow-up Assessment. This is an optional instrument that can be used at the teacher's discretion.

In some units, the questions in the Follow-up Assessment are parallel to those in the Instructional Assessment. (Consult the chart in the introduction to each unit for the specific concepts which each question addresses.) The Follow-up Assessment could be used to check for retention of skills at a later time or as a reassessment after the practice sheets and suggested activities have been completed. It is not necessary to administer the entire instrument. The teacher may wish to use only that portion of the assessment which addresses the concepts with which students had difficulties. Note that the Follow-up Assessment instrument itself does not appear in the teacher materials, but the answer key for it is given in the answer section of each unit.

Cross Reference of Algebridge Modules to Pre-Algebra and Algebra Textbook Topics

	Fractions in Expressions and Equations	Meaning of Negative Numbers	Pattern Recognition and Proportional Reasoning	Constructing Numerical Equations	Attacking Word Problems Successfully	Concept of Variable	Concept of Equality and Inequality	Operations on Equations and Inequalities
Algebraic Expressions	X		X	X	X	X		X
Basic Operations								X
Decimals	X							
Equations and Inequalities	X		X				X	X
Formulas			X	X		X	X	X
Fractions	X		X					
Functions & Graphs			X			X	X	
Graphing Equations and Inequalities							X	
Integers		X						
Integers & Equations		X					X	X
Linear Equations	X		X	X		X	X	X
Linear Inequalities	X		X			X	X	X
Number Properties			X					
Open Sentences					X	X		X
Order of Operations								X
Percent	X		X					
Problem Solving			X	X	X			
Ratio & Proportion	X		X			X	X	
Rational Expressions	X		X					
Rational Numbers	X		X					
Similarity			X					
Solving Equations	X						X	X
Solving Inequalities	X						X	X
Transforming Equations								X
Variables	X		X	X	X	X		
Variation			X			X	X	
Word Problems			X	X	X	X		

ALGE**BRIDGE**™

Fractions

in Expressions

and Equations

Fractions in Expressions and Equations

Introduction

Students often have incomplete understanding or misconceptions about the processes involved in simplifying numerical expressions and equations that involve fractions. The ability to work efficiently with similar algebraic expressions and equations is dependent upon a thorough understanding of these processes. This unit develops, in a numerical context, a student's ability to

- distinguish between expressions and equations and recognize that simplifying expressions involves a different method than that used for clearing equations of fractions;

- recognize that whole numbers and fractions can be expressed in different forms;

- recognize that the number 1 can be expressed in different forms;

- recognize whether a fraction is less than or greater than 1;

- recognize that multiplying or dividing by 1 does not change the value of a number;

- simplify complex fractions;

- eliminate fractions in an equation by multiplying each term in the equation by the least common denominator.

Cross-References to Pre-Algebra and Algebra Textbook Topics	What follows is a compilation of chapter titles and index entries that are found in typical pre-algebra and algebra textbooks. The concepts and abilities addressed in this unit could be used to enhance the instruction on these topics.

decimals	rational expressions
fractions	variables and expressions
percent	solving equations
ratio and proportion	solving inequalities
rational numbers	

List of Terms	The following is a list of words and phrases that are found in this unit.

coefficient	fraction bar
complex fraction	least common denominator
denominator	least common multiple
equal	numerator
equation	prime number
expression	simple fraction
factor	simplify
fraction	term

Classification by Concept

Target Concepts and Abilities	Questions in Instructional Assessment	Questions in Follow-up Assessment	Practice Sheets
Distinguish between expressions and equations	2, 3	1, 2	2, 11
Recognize different names for numbers	1, 5, 6, 8, 11, 17, 18, 19, 20, 21, 22, 24	3, 4, 5, 6, 7, 8, 9, 10, 11, 12	1, 3, 4, 12, 13, 14
Recognize different names for 1	4, 7, 9, 10, 15	13, 14, 15, 16, 17	3, 4, 5, 11, 12, 13
Recognize the multiplicative property of 1	12, 16, 18, 23	9, 19, 20, 21, 22	7, 8, 11, 12, 13, 14, 15
Recognize whether numbers are greater than or less than 1	13, 14	18	6, 7
Simplify complex fractions	23, 24	23, 24	3, 8, 12
Eliminate fractions in equations	25, 26, 27, 28	25, 26, 27, 28	9, 10, 11, 15

Classification by Question

Target Concepts and Abilities		Distinguish between expressions and equations	Recognize different names for numbers	Recognize different names for 1	Recognize the multi-plicative property of 1	Recognize whether numbers are greater than or less than 1	Simplify complex fractions	Eliminate fractions in equations
Question Number or Practice Sheet Number	1	FA	IA PS					
	2	IA FA PS						
	3	IA	FA PS	PS			PS	
	4		FA PS	IA PS				
	5		IA FA	PS				
	6		IA FA			PS		
	7		FA	IA	PS	PS		
	8		IA FA		PS		PS	
	9		FA	IA	FA			PS
	10		FA	IA				PS
	11	PS	IA FA	PS	PS			PS
	12		FA PS	PS	IA PS		PS	
	13		PS	FA PS	PS	IA		
	14		PS	FA	PS	IA		
	15			IA FA	PS			PS
	16			FA	IA			
	17		IA	FA				
	18		IA		IA	FA		
	19		IA		FA			

IA - Questions in Instructional Assessment
FA - Questions in Follow-up Assessment
PS - Practice Sheets

Classification by Question

Target Concepts and Abilities	Distinguish between expressions and equations	Recognize different names for numbers	Recognize different names for 1	Recognize the multiplicative property of 1	Recognize whether numbers are greater than or less than 1	Simplify complex fractions	Eliminate fractions in equations
Question Number or Practice Sheet Number 20		IA		FA			
21		IA		FA			
22		IA		FA			
23				IA		IA FA	
24		IA				IA FA	
25							IA FA
26							IA FA
27							IA FA
28							IA FA

IA - Questions in Instructional Assessment
FA - Questions in Follow-up Assessment
PS - Practice Sheets

Instructional Assessment

The questions in both the Instructional and Follow-up Assessment Instruments for this unit have different formats appropriate to the nature of the concepts addressed in the questions.

It is necessary to introduce students to these formats since they may have limited experience with questions requiring different types of responses.

If there is not sufficient time to review examples of the different question types, review the following directions with the students before they do the *Algebridge*™ exercises.

Directions

There are three types of questions in this exercise.

One type asks you to choose the answer to the problem and mark the corresponding letter, A, B, C, or D, on your answer sheet.

The second type gives you several choices related to a specific question. You must answer **yes** or **no** for each of these choices.

The third type asks you to write your own answer to the problem.

If the Instructional Assessment as a whole seems too long for the students, the questions can be administered in four groups as follows, according to which concepts the teacher wishes to assess.

Questions 1–11 — vocabulary; names for numbers

Questions 12–14 — relative size of numbers

Questions 15–22 — the role of 1 in reducing fractions

Questions 23–28 — complex fractions and equations (least common denominator)

Note: In some questions, a raised dot is used to indicate multiplication. If students are not familiar with this notation, they should be informed that it has the same meaning as the times sign.

Focus of Questions 1–3

Questions 1–3 require the students to demonstrate their understanding of the terms fraction, expression, and equation. It is not as important for students to state a formal definition of these terms as it is for them to be able to recognize the difference between expressions and equations because the rules for manipulating fractions in expressions and equations differ. Question 1, in particular, deals with the term *fraction* for which there are several alternative definitions. Students should not have their alternative answers to questions 1 and 3 marked wrong if they can defend them, but they should be led to understand the rationale for the indicated correct answers.

Question 1

1. Consider each of the following. Is the number represented as a fraction?

yes (no)   a)   2

(yes) no   b) $\dfrac{2}{4}$

yes (no)   c) $3\dfrac{1}{3}$

(yes) no   d) $\dfrac{10}{3}$

(yes) no   e) $\dfrac{2}{100}$

Analysis of Question 1

A fraction, as defined in this unit, is an expression *already* in the form $\frac{a}{b}$ where a and b are expressions and $b \neq 0$. (Another definition of a fraction, not used in this unit, is an expression that *can be put* into the form $\frac{a}{b}$ where $b \neq 0$.) The former definition is more appropriate for this unit because it is consistent with the types of rational expressions that students will encounter in algebra. Also, this definition is consistent with the mathematical manipulations addressed in this unit.

Guiding Class Discussion

The teacher should initiate discussion by asking the students how they would define the word *fraction*. In the course of discussion, the teacher should point out that, although a fraction can be defined in many ways, there are standard rules that govern its use in mathematics. Students should be asked whether 1a and 1c can be expressed as fractions. This discussion should lead to the fact that the number 2 is equivalent to $\frac{4}{2}$ and that $3\frac{1}{3}$ is equivalent to $\frac{10}{3}$. The teacher may wish to include in the discussion of the definitions of a fraction a discussion of the two types of fractions the students will encounter, *i.e.*, proper fractions and improper fractions. The teacher should emphasize the fact that every number can be written in many different forms, and that awareness of this fact is often essential for the solution of a problem that contains fractions.

The practice sheets and suggested activities in the Follow-up Instruction section under the concept **Recognize different names for numbers** may be helpful for those students having difficulty or needing more practice.

Practice sheet 1 can be used at this point to reinforce the definition of fraction as presented in this unit. For this sheet, students are asked to express each given number as a fraction and to write another equivalent fraction.

Question 2

2. Consider each of the following. Is it an equation?

yes (no)   **a)** $\dfrac{2}{3} + 6$

(yes) no   **b)** $\dfrac{2}{3} + 6 = 6\dfrac{2}{3}$

(yes) no   **c)** $5 + \dfrac{4}{5} = \dfrac{29}{5}$

yes (no)   **d)** $\dfrac{7}{3} + \dfrac{8}{3}$

yes (no)   **e)** $\dfrac{4}{5}$

yes (no)   **f)** $5 \div \dfrac{1}{3}$

Analysis of Question 2

An equation, as defined in this unit, is a mathematical statement that contains two expressions with an equal sign between them.

Guiding Class Discussion

Students should be asked to state each part of question 2 (a–f) in words so that the teacher can ascertain whether or not they understand the symbols $+$ and $\div$, as well as $=$. At this time, the teacher may want to discuss with the students statements which use the inequality symbols, $>$, $<$, and $\neq$.

The practice sheets and suggested activities in the Follow-up Instruction section under the concept **Distinguish between expressions and equations** may be helpful for those students having difficulty or needing more practice.

Question 3

3. Consider each of the following. Is it an expression?

(yes) no	**a)**	$6 \div \dfrac{1}{5}$
(yes) no	**b)**	$\dfrac{3}{2} + 4$
yes (no)	**c)**	$\dfrac{3}{2} + 4 = \dfrac{11}{2}$
(yes) no	**d)**	$\dfrac{7}{3} \cdot 4$
(yes) no	**e)**	$6\dfrac{2}{3}$
(yes) no	**f)**	$6 - \dfrac{2}{3}$

Analysis of Question 3

An expression, as defined in this unit, is a symbolic name for a number. In some algebra texts, $6\frac{2}{3}$ (3e) is called a term, not an expression (an expression contains more than one term or requires an operation); in others it is called an expression. The teacher should be aware of the definition used in his or her text when evaluating students' responses to 3e.

Guiding Class Discussion

As in question 2, students should be asked to express a–f in words.

The practice sheets and suggested activities in the Follow-up Instruction section under the concept **Distinguish between expressions and equations** may be helpful for those students having difficulty or needing more practice.

In particular, **practice sheet 2** focuses on this concept by having students match equivalent expressions and form an equation showing the equivalence.

Focus of Question 4

Question 4 emphasizes the fact that the number 1 can be written in many different numerical forms, and that in each case the numerator is equal to the denominator. This concept is fundamental to simplifying fractions.

Question 4

4. Consider each of the following. Is it another name for 1?

    (yes) no   **a)** $\dfrac{7}{7}$

    yes (no)   **b)** $\dfrac{5}{6}$

    (yes) no   **c)** $\dfrac{28}{28}$

    yes (no)   **d)** $\dfrac{10}{9}$

    yes (no)   **e)** $\dfrac{11}{1}$

Guiding Class Discussion

Students who answer 4b, 4d, and 4e incorrectly should be asked to explain how they arrived at their answers. The teacher should ask the students what number 5 should be divided by to get 1. Discussing various representations for division at this point may be helpful for students who do not recognize the fraction bar as a division symbol. Then students could be asked to formulate a rule for determining whether an expression is equal to 1 (*e.g.*, "Any number divided by itself is equal to 1"). If the students then write $\frac{5}{6}$ as $5 \div 6$ and $\frac{11}{1}$ as $11 \div 1$, they can more easily see that these expressions are not equal to 1.

The practice sheets and suggested activities in the Follow-up Instruction section under the concept **Recognize different names for 1** may be helpful for those students having difficulty or needing more practice.

More specifically, **practice sheet 3** is useful for discussing various representations of division while **practice sheets 5 and 6** highlight the idea that if the numerator and denominator of a fraction are equal, then the fraction is another name for 1. Practice sheet 5 can also be used to discuss commutativity of multiplication and addition.

Focus of Question 5

Question 5 assesses whether students understand that the number 12 can be expressed in different forms. This question also assesses whether students understand that multiplying the number 12 by the number 1 does not change the value of the number 12 and that the number 1 itself can be expressed in different forms.

Question 5

5. Consider each of the following. Is it another name for 12?

(yes)   no   **a)** $12 \cdot 1$

(yes)   no   **b)** $12 \cdot \dfrac{7}{7}$

(yes)   no   **c)** $12 \div 1$

(yes)   no   **d)** $12 \div \dfrac{7}{7}$

(yes)   no   **e)** $3 \cdot 4$

yes   (no)   **f)** $\dfrac{3}{4}$

Analysis of Question 5

If students answer 5a and 5c correctly, but answer 5b and 5d incorrectly, they might not realize that $\frac{7}{7}$ is equal to 1.

Parts 5e and 5f illustrate how an operation can play a role in naming the number 12. While both 5e and 5f contain the numbers 3 and 4, the product of these numbers is a correct representation for 12, but the quotient is not.

Guiding Class Discussion

Students may understand that $\frac{3}{3} = 1$, but may not understand that $\frac{213}{213} = 1$. To help the students understand that the number 1 can be written in many ways, the teacher could show and elicit many examples with both large and small numbers.

The practice sheets and suggested activities in the Follow-up Instruction section under the concept **Recognize different names for numbers** may be helpful for those students having difficulty or needing more practice.

Focus of Question 6

Question 6 assesses whether students understand different ways to indicate division as well as whether they can recognize different ways to express the number 15.

Question 6

6. Consider each of the following. Is it another name for 15?

(yes)   no   **a)** $\dfrac{15}{1}$

yes   (no)   **b)** $\dfrac{1}{15}$

yes   no   **c)** $\dfrac{30}{2}$

yes   no   **d)** $30 \div 2$

yes   no   **e)** $2 \div 30$

Analysis of Question 6

If students answer 6a correctly and 6c incorrectly, they should be asked how they arrived at their answers. These students may have divided 30 by 2 incorrectly, or they may not have realized that $\frac{30}{2} = \frac{15}{1} \times \frac{2}{2}$, that is, $\frac{30}{2}$ is the product of $\frac{15}{1}$ and a form of 1, namely $\frac{2}{2}$. Students who answer 6a correctly and 6b incorrectly might not realize that $\frac{15}{1}$ and $\frac{1}{15}$ do not name the same number. If students answer 6c correctly but 6d or 6e incorrectly, they might not understand that $\div$ means "divided by" and that $30 \div 2$ is equivalent to $\frac{30}{2}$, not $\frac{2}{30}$.

Guiding Class Discussion

The practice sheets and suggested activities in the Follow-up Instruction section under the concept **Recognize different names for numbers** may be helpful for those students having difficulty or needing more practice.

     Practice sheet 3 addresses different representations for division and can be used to highlight the equivalence between dividing by n and multiplying by $\frac{1}{n}$. **Practice sheet 14** has a similar focus and could serve as extra practice or review.

Focus of Question 7

Question 7 assesses whether students recognize different ways to express 1 using the four basic operations.

Question 7

7. Consider each of the following. Is it equal to 1?

yes   no   **a)** $\dfrac{1}{3} \div \dfrac{1}{3}$

yes   no   **b)** $\dfrac{1}{3} + \dfrac{1}{3}$

yes   no   **c)** $\dfrac{1}{3} - \dfrac{1}{3}$

yes   no   **d)** $\dfrac{\frac{1}{3}}{\frac{1}{3}}$

yes   no   **e)** $\dfrac{1}{3} \cdot \dfrac{1}{3}$

13

yes⃝ no   **f)** $\dfrac{1}{3} + \dfrac{1}{3} + \dfrac{1}{3}$

yes⃝ no   **g)** $\dfrac{1}{3} \cdot \dfrac{3}{1}$

Analysis of Question 7

Students who answer 7a correctly and 7d incorrectly or vice versa should be helped (*e.g.*, with whole number examples) to realize that

$$\dfrac{\frac{1}{3}}{\frac{1}{3}} = \dfrac{1}{3} \div \dfrac{1}{3}.$$

A good understanding of this equivalence can help students understand later that expressions of the form $\dfrac{\frac{x}{y}}{\frac{x}{y}}$ are equal to 1.

Guiding Class Discussion

The practice sheets and suggested activities in the Follow-up Instruction section under the concept **Recognize different names for 1** may be helpful for those students having difficulty or needing more practice.

Focus of Question 8

Question 8 addresses the issue of preserving the value of a fraction when the fraction is multiplied by 1.

Question 8

8. Consider each of the following. Is it another name for $\dfrac{1}{4}$?

yes⃝ no   **a)** $\dfrac{1}{4} \cdot 1$

yes⃝ no   **b)** $\dfrac{1}{4} \cdot \dfrac{2}{2}$

yes no⃝   **c)** $\dfrac{1}{4} \cdot \dfrac{2}{3}$

yes⃝ no   **d)** $\dfrac{1}{4} \cdot \dfrac{3}{3}$

yes no⃝   **e)** $\dfrac{1}{4} \cdot 4$

Guiding Class Discussion

The teacher should emphasize that the product of any number and 1 preserves the value of that number whether it is a whole number or a fraction, and that the number 1 can have different forms.

Focus of Question 9

The practice sheets and suggested activities in the Follow-up Instruction section under the concept **Recognize different names for numbers** may be helpful for those students having difficulty or needing more practice.

Question 9 deals with recognizing different names for 1 with a focus on the use of operations in expressions.

Question 9

9. Consider each of the following. Is it another name for 1?

(yes) no	**a)**	$\dfrac{7+2}{7+2}$
(yes) no	**b)**	$\dfrac{7+2}{2+7}$
yes (no)	**c)**	$\dfrac{7-2}{2-7}$
(yes) no	**d)**	$\dfrac{2 \cdot 7}{2 \cdot 7}$
(yes) no	**e)**	$\dfrac{7 \cdot 2}{2 \cdot 7}$
yes (no)	**f)**	$\dfrac{7 \div 2}{2 \div 7}$

Analysis of Question 9

Algebra students frequently have trouble understanding that $\frac{x+y}{x+y} = 1$, but $\frac{x-y}{y-x} = -1$. Because the same numbers, 7 and 2, are used in each part of this question, the student must concentrate on the operations involved in order to determine the value of each expression.

Guiding Class Discussion

By replacing the numbers 7 and 2 in question 9, first with larger numbers and then with placeholders, the teacher can ascertain whether students understand how to manipulate similar fractions. Having students compare the terms in the expressions $\frac{7-2}{2-7}$ by writing $7 - 2 = -(2 - 7)$ might help them understand that the numerator and denominator of a fraction must be equal in order for a fraction to be equal to 1.

The practice sheets and suggested activities in the Follow-up Instruction section under the concept **Recognize different names for 1** may be helpful for those students having difficulty or needing more practice.

Focus of Questions 10–11   In questions 10 and 11, students must write different names for a number, with the option of using operations in expressions.

Question 10

10. Write four different expressions (names) for the number 1.

Answer: $2 \times \frac{1}{2}$ $4 \div 4$

$4 - 3$ $\frac{1}{2} + \frac{1}{2}$

Analysis of Question 10   The evaluation of students' answers to question 10 will help the teacher determine whether the students understand the concept addressed in question 4, 7, and 9. If there is little or no diversity in the form of the answer, the teacher may want to question the student further to be sure the concept of many names for 1 has been fully understood.

Guiding Class Discussion   The practice sheets and suggested activities in the Follow-up Instruction section under the concept **Recognize different names for 1** may be helpful for those students having difficulty or needing more practice.

Question 11

11. Write four different expressions (names) for the number $\frac{2}{3}$.

Answer: $\frac{2}{3} \times 1$ $\frac{2}{1} \times \frac{1}{3}$

$\frac{3}{3} - \frac{1}{3}$ $\frac{4}{6}$

Guiding Class Discussion   Question 11 gives the teacher the opportunity to evaluate the students' understanding of equivalence of fractions. Again, the discussion could be stimulated by having students evaluate each other's answers.

The practice sheets and suggested activities in the Follow-up Instruction section under the concept **Recognize different names for numbers** may be helpful for those students having difficulty or needing more practice.

In particular, **practice sheet 4** presents questions similar to question 11. For this sheet, students write expressions for chosen numbers using one or more of the basic operations.

Focus of Question 12   Question 12 assesses whether students recognize the multiplicative property of 1. It requires that they understand the concepts presented in questions 4–7.

Question 12

12. Consider each of the following. Is it a true statement?

yes **(no)**   **a)** $\dfrac{3}{2} \times \dfrac{2}{3} = \dfrac{2}{3}$

(yes) no   **b)** $\dfrac{3}{3} \times \dfrac{2}{3} = \dfrac{2}{3}$

(yes) no   **c)** $\dfrac{2}{2} \times \dfrac{2}{3} = \dfrac{2}{3}$

yes **(no)**   **d)** $3 \times \dfrac{2}{3} = \dfrac{2}{3}$

yes **(no)**   **e)** $6 \times \dfrac{2}{3} = \dfrac{2}{3}$

Analysis of Question 12

If students answer any part of question 12 incorrectly, the teacher should ask them to explain how they arrived at their answers to determine if students are making errors in using the multiplication algorithm for fractions rather than looking for representations of 1. A review of questions 4–7 may be helpful.

Guiding Class Discussion

The practice sheets and suggested activities in the Follow-up Instruction section under the concept **Recognize the multiplicative property of 1** may be helpful for those students having difficulty or needing more practice.

Focus of Question 13

Question 13 assesses the students' ability to recognize whether a number is greater than 1 when it is expressed as a fraction by comparing the relative sizes of the numerator and denominator.

Question 13

13. Which of the following numbers is greater than 1?

(A) $\dfrac{3}{4}$

(B) $\dfrac{4}{3}$

(C) $\dfrac{4}{4}$

(D) $\dfrac{3}{3}$

Guiding Class Discussion

Students should be asked to give reasons for their answers. The discussion should lead to the conclusion by the students that the relative size of the numerator and denominator determines the size of the fraction with respect to 1. The ability to judge the value of a fraction relative to 1 is helpful in the development of estimation skills for use in solving problems that involve fractions.

The practice sheets and suggested activities in the Follow-up Instruction section under the concept **Recognize whether numbers are greater than or less than 1** may be helpful for those students having difficulty or needing more practice.

Practice sheet 6 would be particularly appropriate at this point. For this sheet, students decide whether the given fractions are greater than, equal to, or less than 1. **Practice sheet 5** also addresses this idea.

Focus of Question 14

Question 14 extends the idea that multiplying a number by 1 does not change its value to the idea that multiplying a number by a number less than (or greater than) 1 yields a value that is less than (or greater than) the original number.

Question 14

14. Consider each of the following. Is it a true statement? (Remember that "□ > △" means "□ is greater than △," and "□ < △" means "□ is less than △.")

   (yes) no   **a)** $12 \cdot \dfrac{4}{3} > 12$

   (yes) no   **b)** $12 \cdot \dfrac{3}{4} < 12$

   yes (no)   **c)** $12 \cdot \dfrac{4}{4} > 12$

   yes (no)   **d)** $12 \cdot \dfrac{3}{3} < 12$

Analysis of Question 14

Students should be shown that, in order to answer a question such as this, it is not necessary for them to calculate the value to the left of the inequality symbol. Instead, the answer can be found by determining whether the fraction by which 12 is being multiplied is less than, greater than, or equal to 1.

Guiding Class Discussion

The practice sheets and suggested activities in the Follow-up Instruction section under the concept **Recognize whether numbers are greater than or less than 1** may be helpful for those students having difficulty or needing more practice.

More specifically, for an exercise similar to question 12, see **practice sheet 7**. For this sheet, students must decide if the given expression is greater than, equal to, or less than 17.

Focus of Questions 15–17

The focus in questions 15–17 is on the role of 1 in the simplification of fractions. Question 17 uses the concepts assessed in questions 15 and 16 to simplify the expression.

Question 15

15. Is $\dfrac{3 \cdot 2 \cdot 5 \cdot 7}{3 \cdot 7 \cdot 11}$ equal to 1?

    (A)   Yes
    (B)   No

Analysis of Question 15

The teacher should point out to the students that it is not necessary to compute the value of the numerator and denominator in order to decide whether an expression in factored form is equal to 1. It is only necessary to determine whether it is possible to match each factor in the numerator with an identical factor in the denominator, that is, to find different forms of 1.

Guiding Class Discussion

The practice sheets and suggested activities in the Follow-up Instruction section under the concept **Recognize different names for 1** may be helpful for those students having difficulty or needing more practice.

Question 16

16. Consider each of the following. Is it equal to $\dfrac{3 \cdot 2 \cdot 5 \cdot 7}{3 \cdot 7 \cdot 11}$?

    (yes)   no   **a)** $\dfrac{2 \cdot 5}{11}$

    yes   (no)   **b)** $\dfrac{2 \cdot 5}{7 \cdot 11}$

    yes   (no)   **c)** $\dfrac{3 \cdot 7}{3 \cdot 7}$

    (yes)   no   **d)** $\dfrac{2 \cdot 5 \cdot 7}{7 \cdot 11}$

Analysis of Question 16

As shown in question 15, finding different forms of 1 in expressions such as those in question 16 (*i.e.*, $\frac{3}{3}$ and $\frac{7}{7}$) and expressing large numbers in factored form are methods of simplifying fractions. These techniques can lead to a better understanding of simplifying algebraic expressions such as $\dfrac{a^2bc}{b^2cd}$.

Guiding Class Discussion

Discussing the simplification of fractions by factoring may help students who answer any part of question 16 incorrectly. The teacher should reinforce the idea of many names for 1 by having students "match up" factors in numerators and denominators to make 1, for example,

$$\frac{210}{231} = \frac{21 \cdot 10}{77 \cdot 3} = \frac{3 \cdot 7 \cdot 2 \cdot 5}{7 \cdot 11 \cdot 3} = \frac{(3) \cdot (7) \cdot 2 \cdot 5}{(3) \cdot (7) \cdot 11} = \frac{10}{11}.$$

This activity also reinforces the concepts of prime numbers and divisibility.

The practice sheets and suggested activities in the Follow-up Instruction section under the concept **Recognize the multiplicative property of 1** may be helpful for those students having difficulty or needing more practice.

Question 17

17. $\dfrac{3 \cdot 2 \cdot 5 \cdot 7}{3 \cdot 7 \cdot 11} =$

(A) $\dfrac{10}{21}$

(B) $\dfrac{5}{7}$

(C) $\dfrac{10}{11}$

(D) 10

Analysis of Question 17

In question 17, students simplify the expression that was examined in questions 15 and 16. Students who answer incorrectly should be encouraged to "match up" factors in the numerator and denominator to make 1.

Guiding Class Discussion

The practice sheets and suggested activities in the Follow-up Instruction section under the concept **Recognize different names for numbers** may be helpful for those students having difficulty or needing more practice.

Focus of Questions 18–20

Questions 18–20 form a set that should be used to help students understand the process they are using when they reduce fractions. A common misconception among students is that $\frac{x+a}{y+a}$ is equal to $\frac{x}{y}$ when $a \neq 0$. In question 18, the numbers are small enough so the students can do the arithmetic to determine whether the equality exists; however, they should be encouraged to answer the question by inspection without calculating the value of the numerator and denominator. The teacher should use questions 19 and 20 to demonstrate the generalization of the methods that can be used to simplify (reduce) fractions.

Question 18

18. Consider each of the following. Is it equal to $\dfrac{5}{7}$?

     yes   (no)   **a)** $\dfrac{5+2}{7+2}$

     (yes)   no   **b)** $\dfrac{5 \cdot 2}{7 \cdot 2}$

     (yes)   no   **c)** $\dfrac{2 \cdot 5}{7 \cdot 2}$

     (yes)   no   **d)** $\dfrac{5(4+2)}{7(4+2)}$

     yes   (no)   **e)** $\dfrac{5-2}{7-2}$

Guiding Class Discussion

To encourage the use of inspection rather than calculation, the teacher should focus on the effect of adding or subtracting numbers to or from the numerator and/or denominator of a fraction. Students should reach the conclusion that adding or subtracting 0 does not change the value of the fraction, but adding a number other than 0 does change the value. For example, $\frac{5+0}{7+0}$ *is equal to* $\frac{5}{7}$, whereas $\frac{5+2}{7+2}$ *is not* equal to $\frac{5}{7}$. Students should recognize that choices b and c contain other names for 1 and thus are easily evaluated by inspection.

The practice sheets and suggested activities in the Follow-up Instruction section under the concepts **Recognize different names for number** and **Recognize the multiplicative property of 1** may be helpful for those students having difficulty or needing more practice.

Question 19

19. Consider each of the following. Is it equal to $\dfrac{93}{72}$?

     (yes)   no   **a)** $\dfrac{93 \cdot 58}{72 \cdot 58}$

     (yes)   no   **b)** $\dfrac{58 \cdot 93}{72 \cdot 58}$

     yes   (no)   **c)** $\dfrac{93 + 58}{72 + 58}$

     yes   (no)   **d)** $\dfrac{93 - 58}{72 - 58}$

     (yes)   no   **e)** $\dfrac{93(61 + 58)}{72(61 + 58)}$

Guiding Class Discussion

To help students make the transition from question 18 to question 20, the teacher should ask them to compare questions 18 and 19 and to note any patterns. If students answer questions 18 and 19 correctly, but have trouble with question 20, it may be helpful to show them how to use question 18 or 19 as a model for answering question 20, with question 19 as a transition problem in which calculation is more difficult and seeing the pattern more important.

The practice sheets and suggested activities in the Follow-up Instruction section under the concept **Recognize different names for numbers** may be helpful for those students having difficulty or needing more practice.

Question 20

20. Consider each of the following. If a number is placed in the □ and a different number is placed in the △, is the expression equal to $\dfrac{\square}{\triangle}$?

yes (no)   **a)** $\dfrac{\square + 2}{\triangle + 2}$

yes (no)   **b)** $\dfrac{\square - 2}{\triangle - 2}$

(yes) no   **c)** $\dfrac{\square(4+2)}{\triangle(4+2)}$

(yes) no   **d)** $\dfrac{2 \cdot \square}{\triangle \cdot 2}$

(yes) no   **e)** $\dfrac{\square \cdot 2}{\triangle \cdot 2}$

Guiding Class Discussion

In the discussion the teacher should focus student attention on other names for 1 and adding and subtracting 0. However, since placeholders are introduced in question 20, the teacher may want to discuss division by zero with the students by posing the question of what would happen if △ were replaced by 2 in 20b or 0 in 20c, d, or e. Students should be made aware of why division by zero is said to be undefined.

To help students understand what values should be excluded from a set of possible replacements for a placeholder in an expression similar to 20b, the teacher could present other fractions with placeholders in the denominators and ask the students what values would make the denominators equal to 0.

The practice sheets and suggested activities in the Follow-up Instruction section under the concept **Recognize different names for**

numbers may be helpful for those students having difficulty or needing more practice.

In particular, for more practice with this concept using placeholders, see **practice sheet 12**. This sheet also extends the concept by addressing the role of other names for 1 in the addition of fractions.

Focus of Questions 21–22

Questions 21 and 22 serve to summarize the concepts in questions 18–20.

Questions 21–22

21. What number can you place in the $\triangle$ to make $\dfrac{5 - \triangle}{7 - \triangle} = \dfrac{5}{7}$?

Answer: *0*

22. What number can you place in the $\square$ to make $\dfrac{5 \cdot \square}{7 \cdot \square} = \dfrac{5}{7}$?

Answer: *Any nonzero number*

Guiding Class Discussion

Questions 21 and 22 should be discussed together. The teacher should ask students why there is only one answer to question 21 but many answers to question 22. Have the students read, describe in words, and explain the following statements:

(1) $\quad \dfrac{5 \cdot \square}{7 \cdot \square} = \dfrac{5}{7} \cdot \dfrac{\square}{\square} = \dfrac{5}{7}$

(2) $\quad \dfrac{5 - \square}{7 - \square} \neq \dfrac{5}{7} - \dfrac{\square}{\square}$

The practice sheets and suggested activities in the Follow-up Instruction section under the concept **Recognize different names for numbers** may be helpful for those students having difficulty or needing more practice.

Practice sheet 13 can be used at this point to review ideas about different names for numbers and the multiplicative property of 1. For this sheet, students decide whether the given expressions are equivalent and write an explanation justifying their decision.

Focus of Question 23

Question 23 illustrates that, no matter how complex the fraction, its value is not changed when it is multiplied by 1 (in any form).

Question 23

23. Consider each of the following expressions. Is it equal to $\dfrac{\frac{2}{3}}{\frac{1}{4}}$?

(yes)   no   **a)** $\dfrac{\frac{2}{3}}{\frac{1}{4}} \cdot \dfrac{3}{3}$

(yes)   no   **b)** $\dfrac{\frac{2}{3}}{\frac{1}{4}} \cdot \dfrac{12}{12}$

(yes)   no   **c)** $\dfrac{\frac{2}{3}}{\frac{1}{4}} \cdot \dfrac{4}{4}$

(yes)   no   **d)** $\dfrac{\frac{2}{3}}{\frac{1}{4}} \cdot \dfrac{7}{7}$

(yes)   no   **e)** $\dfrac{\frac{2}{3}}{\frac{1}{4}} \cdot \dfrac{24}{24}$

yes   (no)   **f)** $\dfrac{\frac{2}{3}}{\frac{1}{4}} \cdot \dfrac{4}{3}$

yes   (no)   **g)** $\dfrac{\frac{2}{3}}{\frac{1}{4}} \cdot 12$

Guiding Class Discussion

The teacher might focus discussion on the various correct answer choices and their implications for simplifying the complex fraction, using this as an opportunity to mention the least common denominator. Students need to be made aware that, although 23b may be the best choice to use to *simplify* the complex fraction, the question asks whether the expression equals $\dfrac{\frac{2}{3}}{\frac{1}{4}}$. They should recognize that, as long as the complex fraction is multiplied by 1, its value is retained.

The practice sheets and suggested activities in the Follow-up Instruction section under the concepts **Recognize the multiplicative property of 1** and **Simplify complex fractions** may be helpful for those students having difficulty or needing more practice.

To focus specifically on simplifying complex fractions, see **practice sheet 8**. Several exercises on **practice sheet 12** also involve complex fractions.

Focus of Question 24

Question 24 introduces the process of simplifying complex fractions by focusing on the fact that dividing by n is the same as multiplying by $\frac{1}{n}$.

Question 24

24. Consider each of the following. Is it equal to $\dfrac{\frac{3}{5}}{2}$?

yes (no) **a)** $\dfrac{3}{5} \cdot 2$

(yes) no **b)** $\dfrac{3}{5} \cdot \dfrac{1}{2}$

(yes) no **c)** $\dfrac{3}{5} \div 2$

yes (no) **d)** $\dfrac{\frac{3}{5} \cdot 5}{2}$

(yes) no **e)** $\dfrac{\frac{3}{5} \cdot 5}{2 \cdot 5}$

Analysis of Question 24

Question 24b and 24c call attention to the equivalence of division by 2 and multiplication by $\frac{1}{2}$. Students often have trouble understanding this concept so it should get special attention. Students may need to be reminded that $2 = \frac{2}{1}$ and not $\frac{1}{2}$. Question 24 also presents the students with another type of complex fraction to simplify. It may be helpful to have students write $\dfrac{\frac{3}{5}}{2} = \dfrac{\frac{3}{5}}{\frac{2}{1}}$ if they are confused about how to answer this question, even though they answered question 23 correctly. Question 24e illustrates the roles of the number 1 in simplifying complex fractions.

Guiding Class Discussion

The practice sheets and suggested activities in the Follow-up Instruction section under the concepts **Recognize different names for numbers** and **Simplify complex fractions** may be helpful for those students having difficulty or needing more practice.

Focus of Questions 25–27

Questions 25–27 form a set of three questions in which students are given an equation and asked to find a common denominator for the fractions in the equation.

Question 25

25. Consider each of the following. Would the result be an equation without fractions?

yes ⟨no⟩   **a)**   Multiply each term in $4 + \dfrac{5}{7} = \dfrac{33}{7}$ by 4.

yes ⟨no⟩   **b)**   Multiply each term in $4 + \dfrac{5}{7} = \dfrac{33}{7}$ by 5.

⟨yes⟩ no   **c)**   Multiply each term in $4 + \dfrac{5}{7} = \dfrac{33}{7}$ by 7.

⟨yes⟩ no   **d)**   Multiply each term in $4 + \dfrac{5}{7} = \dfrac{33}{7}$ by 14.

yes ⟨no⟩   **e)**   Multiply each term in $4 + \dfrac{5}{7} = \dfrac{33}{7}$ by 20.

Guiding Class Discussion

Question 25 introduces the concept of eliminating fractions from an equation in order to make it easier to solve. Since multiplying the terms in the equation by either 7 or 14 achieves the desired result, students can also see that the least common denominator of the fractions is *not* the *only* number that can be used to remove the fractions. It will, however, yield the smallest numbers (which may be easier to work with). The teacher should point out to the students that multiplying $\frac{5}{7}$ and all the other terms in an equation by 7 is permissible because equality is maintained after the operation has been performed. But multiplying the expression $\frac{5}{7}$ by 7 to simplify $\frac{5}{7}$ is *not* permissible because, in this case, the value of $\frac{5}{7}$ will be changed by the multiplication.

Using a double-pan scale or other balancing device to demonstrate the idea of maintaining balance in an equation may help students understand this distinction. It may be helpful, also, to remind students that one should multiply a number by 1 only when the objective is to preserve the value of that number.

The practice sheets and suggested activities in the Follow-up Instruction section under the concept **Eliminate fractions in equations** may be helpful for those students having difficulty or needing more practice.

Practice sheets 9 and 10 are particularly appropriate for use with questions 25–28.

Question 26

26. Consider each of the following. Would the result be an equation without fractions?

yes (no)   **a)**   Multiply each term in $\dfrac{5}{2} + 3 + \dfrac{7}{3} = \dfrac{47}{6}$ by 2.

yes (no)   **b)**   Multiply each term in $\dfrac{5}{2} + 3 + \dfrac{7}{3} = \dfrac{47}{6}$ by 3.

(yes) no   **c)**   Multiply each term in $\dfrac{5}{2} + 3 + \dfrac{7}{3} = \dfrac{47}{6}$ by 6.

yes (no)   **d)**   Multiply each term in $\dfrac{5}{2} + 3 + \dfrac{7}{3} = \dfrac{47}{6}$ by 11.

(yes) no   **e)**   Multiply each term in $\dfrac{5}{2} + 3 + \dfrac{7}{3} = \dfrac{47}{6}$ by 12.

Analysis of Question 26

Students who answer **yes** to 26d may be adding the denominators, especially if they answer **no** to 26e. It might be helpful to ask them to write what the equation would look like after all the terms were multiplied by 11. Students who answer **yes** to 26a and 26b may argue that multiplying by 2, then by 3, achieves the desired result. The efficiency of this procedure, as well as the insufficiency of multiplying by just one of the two, could then be discussed. In discussing question 26 with students, the teacher should ask whether there are any other numbers that would work.

Guiding Class Discussion

The practice sheets and suggested activities in the Follow-up Instruction section under the concept **Eliminate fractions in equations** may be helpful for those students having difficulty or needing more practice.

Question 27

27. Consider each of the following numbers. If each term in the equation

$$\frac{10}{3} + 1 + \frac{8}{9} = \frac{21}{5} + \frac{2}{9} + \frac{4}{5}$$

were multiplied by the number, would the result be an equation without fractions?

yes (no)   **a)**   3
yes (no)   **b)**   5
yes (no)   **c)**   9
yes (no)   **d)**   12
yes (no)   **e)**   14
(yes) no   **f)**   45
(yes) no   **g)**   135

Analysis of Question 27

Question 27 is the third in the set of equations in which students are asked to find a common denominator. This time a correct multiplier does not appear in the equation. The teacher should ask the students who answered 27f correctly how they arrived at their answer.

Guiding Class Discussion

It may be helpful to show students that factoring may help to find the *least* common denominator (multiple): $3 \cdot 9 \cdot 5 = 3 \cdot (3 \cdot 3) \cdot 5$. Since 3 is a factor of 9, it does not have to be included. Therefore, the least common denominator (multiple) is $9 \cdot 5$.

The practice sheets and suggested activities in the Follow-up Instruction section under the concept **Eliminate fractions in equations** may be helpful for those students having difficulty or needing more practice.

In particular, for more practice with denominators in factored form, the teacher may want to use **practice sheet 10**.

Focus of Question 28

Question 28 assesses whether the students are able to determine a number that can be used as a multiplier to clear an equation of fractions and to use that number to write a correct equation without fractions.

Question 28

28. Multiply each term of $\dfrac{9}{5} + \dfrac{7}{10} = \dfrac{25}{10}$ by some number to get an equation without fractions.

    a)   Write your equation.

        Answer: *For example: 18 + 7 = 25*

    b)   What number did you use?

        Answer: *10*

Analysis of Question 28

A combination of answers such as $9 + 7 = 25$ to the first question and 10 to the second would indicate that the student could choose a correct multiplier but does not understand what to do with it after it is chosen. The teacher may have to work through several similar examples before students fully understand the role of the multiplier.

Guiding Class Discussion

The practice sheets and suggested activities in the Follow-up Instruction section under the concept **Eliminate fractions in equations** may be helpful for those students having difficulty or needing more practice.

Note that in addition to **practice sheets 9 and 10**, mentioned previously, **practice sheets 11 and 15** also address this concept. The teacher may find practice sheet 11 particularly useful for clarifying students' conceptions since this sheet focuses on the differences between *simplifying* an expression involving fractions and *clearing* an equation of fractions.

Follow-up Instruction

After the Instructional Assessment has been used to pinpoint students' conceptual weaknesses, the teacher may wish to use the following suggested activities and practice sheets to help correct those misconceptions.

Distinguish between expressions and equations

Activity. The teacher could make a set of flash cards with a numerical representation on one side, the reverse side indicating whether it is an expression or an equation. (Note: The teacher may want to have students make their own cards and then pair up to test each other.)

Alternatively, the student could be presented with a set of cards that have only numerical representations on one side (with no identification on the reverse side). The student could then be instructed to separate the set into two piles—expressions and equations.

For a third activity, students could be given sets of flash cards with numbers, signs of operations, and equal signs, and asked to arrange the cards to form equations and expressions.

On **practice sheet 2** students compare equivalent expressions and write equations using the expressions. This sheet also can be used to review basic operations with fractions.

Practice sheet 11 requires the students to compare the steps taken in adding fractions with those taken in clearing the fractions from a numerical equation.

Questions 1 and 2 in the Follow-up Assessment also address this concept.

Recognize different names for numbers

Activity. The teacher could make a set of flash cards with a numerical representation on one side, the reverse side indicating whether it is a fraction or not a fraction. (Note: The teacher may want to have students make the cards and then pair up to test each other.)

As an alternate activity, the teacher could present the student with a set of cards that have only numerical representations on one side (with no identification on the reverse side) and instruct the student to separate the set into two piles—fractions and forms of numbers that are not fractions.

Activity. As a follow-up to question 5, students could be asked to find additional ways to express 12 using only 3's and/or 4's.

Activity. As a follow-up activity to question 20, the teacher might ask students to graph the function $\frac{5}{\triangle}$ for $\triangle = 1, 2, 3, 4, 5, 6, 7, 8, 10$. Then have students vary the numerator and graph $\frac{2}{\triangle}$ and $\frac{10}{\triangle}$, for example, using a different color pen for each numerator. This activity helps the students see what happens to the value of a fraction as the denominator or numerator increases or decreases.

On **practice sheet 1** students write numbers in fractional form. This sheet can be used to focus on the definition of fraction presented in this unit and to emphasize the equivalence between various fractional forms of a number and other representations of the same number.

On **practice sheet 3** students rewrite fractions using other representations for division, recognizing that dividing by n is the same as multiplying by $\frac{1}{n}$.

Practice sheet 4 specifically emphasizes that by using one or more of the basic operations, numbers can be written in many different forms.

Questions 3–12 in the Follow-up Assessment also address this concept.

Recognize different names for 1

Activity. To help students understand the concept of different expressions for 1, they can evaluate their classmates' answers to question 10 by exchanging papers or putting their answers on the board for discussion. As another option, the teacher could write some of the responses on a transparency or a worksheet for students to evaluate.

For a group or individual, the teacher could have students use any number of the following cards to form expressions equal to 1:

$$\boxed{3} ; \boxed{3} ; \boxed{6} ; \boxed{6} ; \boxed{\div} ; \boxed{+} ; \boxed{(} ; \boxed{)} .$$

Those expressions could include: $3 \div 3$; $6 \div 6$; $(3 + 3) \div 6$; $6 \div (3 + 3)$; $36 \div 36$; $63 \div 63$.

For another activity, give students the following cards:

$$\boxed{\tfrac{4}{5}} ; \boxed{\tfrac{5}{4}} ; \boxed{\times} ; \boxed{\div} .$$

The purpose of this activity is to use the cards to form three different expressions that are equal to 1, for example, $\frac{4}{5} \div \frac{4}{5}$.

On **practice sheet 3** students rewrite fractions using other representations for division. This helps them to recognize that dividing by n is the same as multiplying by $\frac{1}{n}$ and to understand that when the numerator and denominator of a fraction are the same, the fraction is another name for 1.

On **practice sheet 4** students are asked to write numbers in different forms. The teacher could ask the students specifically to include different names for one in the various expressions for their numbers.

Practice sheet 5 instructs students to compare the numerator and denominator of a fraction and then decide whether the fraction is equal to or not equal to 1.

Practice sheet 11 requires the students to compare the steps taken in adding fractions with those taken in solving a numerical equation that contains fractions. Students must use other names for 1 in order to express each fraction with a common denominator.

Questions 13–17 in the Follow-up Assessment address this concept.

Recognize the multiplicative property of 1

Activity. If students are familiar with calculators and with decimals, they can play a "target" game where a number, say 24, is taken as the target. The first player chooses a number, say 17, and the second tries to find a multiplier so that when the product of 17 and the number is found on the calculator, the display shows 24 followed by zeros in at least the first two places. If the target is not hit, the next player chooses a multiplier to use with the displayed number. For example, the first multiplier chosen is 1.5; the displayed number is 25.5 (17×1.5). The second multiplier chosen is 0.9; the displayed number is 22.95. The third multiplier chosen is 1.1; the displayed number is 25.245; and so on. Students quickly learn the effect of multipliers just above and just below 1.

On **practice sheet 7** students determine whether an expression is greater than, equal to, or less than a given number. The teacher may want to have the students explain why calculation is not necessary for this exercise. Students should be able to explain that knowing whether the given number (or factor by which 17 is being multiplied) is less than, equal to, or greater than 1 determines whether the result will be, respectively, less than, equal to, or greater than 17.

Practice sheet 8 has students apply the multiplicative property of 1 to simplify complex fractions.

Practice sheet 11 requires the students to compare the steps taken in adding fractions with those taken in solving a numerical equation that contains fractions.

Questions 9 and 19–22 in the Follow-up Assessment address this concept.

Recognize whether numbers are greater than or less than 1

Activity. Those students who are having difficulties determining whether a fraction is less than, equal to, or greater than 1 may benefit from using rods or fraction bars to model the fractions. The manipulatives may make it easier for students to focus on the relative sizes of the numerator and the denominator. Students may work in pairs with one student constructing the fraction and the other student determining if it is less than, equal to, or greater than 1.

On **practice sheet 6** students are given a list of fractions and asked whether each fraction is less than, equal to, or greater than 1. Students apply this concept on **practice sheet 7** when they determine whether an expression is greater than, equal to, or less than a given number.

Question 18 in the Follow-up Assessment also addresses this concept.

Simplify complex fractions

Activity. Since many students have difficulty understanding that division by n is the same as multiplication by $\frac{1}{n}$, the teacher may want to demonstrate this equivalence by using manipulatives. For example, to show that $10 \div 2$ is the same as $10 \times \frac{1}{2}$, separate 10 toothpicks into groups of 2. The result is 5 groups. Taking $\frac{1}{2}$ of a group of 10 toothpicks also yields a result of 5. Encourage students to model similar problems using toothpicks or other materials until they are convinced of the equivalence.

On **practice sheet 3** students rewrite fractions using other representations for division. The students could model the problems on this sheet in the manner described in the previous activity to emphasize that division by n is the same as multiplication by $\frac{1}{n}$.

Practice sheet 8 has students simplify complex fractions.

Questions 23 and 24 in the Follow-up Assessment also address this concept.

Eliminate fractions in equations

On **practice sheets 9 and 10** students use a common denominator to clear an equation of fractions. Some of the questions on practice sheet 10 have denominators that are in factored form. Consequently, the teacher may use this sheet to focus a discussion on the use of prime factorization to find the LCD.

Practice sheet 11 requires the students to compare the steps taken in adding fractions with those taken in solving a numerical equation that contains fractions.

Questions 25–28 in the Follow-up Assessment also address this concept.

All unit concepts

Practice Sheets 12–15 require students to apply several concepts at one time and thus can be used during instruction as a summary or review of the unit prior to the administration of the Follow-up Assessment. The format of the questions is similar to the Assessment questions but also may require that the student *explain* why a choice is or is not correct. The student's explanation can be used to determine any remaining misunderstandings about the concepts presented.

Practice sheet 12 presents several of the concepts that deal with the equivalence of fractions including recognizing different names for numbers, recognizing different names for 1, and recognizing the multiplicative property of 1. The questions will help the teacher determine if students have conflicting ideas about working with fractions in both expressions and equations.

Practice sheet 13 also deals with recognizing different names for numbers and the multiplicative property of 1. Rather than calculating the values in each column, students should use the number patterns and number properties that they know to decide whether the expressions are equal. The questions also present opportunities to discuss procedures for adding and subtracting fractions.

Practice sheet 14 emphasizes that a fraction is an indicated division and that the fraction can be written using other symbols for division. Students need to be aware that the order in which the numbers and symbols are written is important. They also must recognize other names for 1 and apply the multiplicative property of 1.

Practice sheet 15 demonstrates that more than one multiplier will yield an equation without fractions. It also reminds students about some properties of zero. Question 2 focuses on the types of errors that students may make when eliminating fractions from equations.

Bibliography of Resource Materials

Arithmetic Teacher, Focus issue: Rational Numbers 31 (February 1984).

Bennett, Albert, and Patricia Davidson. *Fraction Bars.* Fort Collins, CO: Scott Resources, 1973.

Easterday, Kenneth E., Loren L. Henry, and F. Morgan Simpson, eds. *Activities for Junior High School and Middle School Mathematics: Readings from the Arithmetic Teacher and the Mathematics Teacher.* Reston, VA: National Council of Teachers of Mathematics, 1981.

Greenwood, Jay. *Developing Mathematical Thinking: Fractions.* Portland, OR: Multnomah ESD, 1986.

Maletsky, Evan, and Christian Hirsch, eds. *Activities from the Mathematics Teacher.* Reston, VA: National Council of Teachers of Mathematics, 1981.

Answers

Answers to Instructional Assessment

1. a) no
b) yes
c) no
d) yes
e) yes
2. a) no
b) yes
c) yes
d) no
e) no
f) no
3. a) yes
b) yes
c) no
d) yes
e) yes
f) yes
4. a) yes
b) no
c) yes
d) no
e) no

5. a) yes
b) yes
c) yes
d) yes
e) yes
f) no
6. a) yes
b) no
c) yes
d) yes
e) no
7. a) yes
b) no
c) no
d) yes
e) no
f) yes
g) yes
8. a) yes
b) yes
c) no
d) yes
e) no

9. a) yes
b) yes
c) no
d) yes
e) yes
f) no
10. Answers will vary. For example,
$2 \times \frac{1}{2}$
$4 \div 4$
$4 - 3$
$\frac{1}{2} + \frac{1}{2}$
11. Answers will vary. For example,
$\frac{2}{3} \times 1$
$\frac{2}{1} \times \frac{1}{3}$
$\frac{3}{3} - \frac{1}{3}$
$\frac{4}{6}$

12. a) no
b) yes
c) yes
d) no
e) no
13. B
14. a) yes
b) yes
c) no
d) no
15. B
16. a) yes
b) no
c) no
d) yes
17. C
18. a) no
b) yes
c) yes
d) yes
e) no
19. a) yes
b) yes

c) no
d) no
e) yes
20. a) no
b) no
c) yes
d) yes
e) yes
21. 0
22. Any nonzero number
23. a) yes
b) yes
c) yes
d) yes
e) yes
f) no
g) no
24. a) no
b) yes
c) yes

d) no
e) yes
25. a) no
b) no
c) yes
d) yes
e) no
26. a) no
b) no
c) yes
d) no
e) yes
27. a) no
b) no
c) no
d) no
e) no
f) yes
g) yes
28. For example,
a) $18+7 = 25$
b) 10

Answers to Practice Sheets

Practice sheet 1

Answers will vary. For example,

1. $\frac{6}{10}$, $\frac{3}{5}$
2. $\frac{11}{2}$, $\frac{22}{4}$
3. $\frac{3}{6}$, $\frac{1}{2}$
4. $\frac{1}{1}$, $\frac{2}{2}$
5. $\frac{8}{9}$, $\frac{16}{18}$
6. $\frac{18}{1}$, $\frac{36}{2}$
7. $\frac{75}{100}$, $\frac{3}{4}$
8. $\frac{3}{100}$, $\frac{6}{200}$

Practice sheet 2

1. $2\frac{1}{2} = 2 + \frac{1}{2}$
2. $22 = 11 + 11$
3. $2 + 2 + 2 = 6$
4. $2 \cdot (2 + 2) = 2 \cdot 2 + 2 \cdot 2$
5. $\frac{1}{2} + \frac{1}{2} = \frac{2}{2}$
6. $[\frac{1}{2} \div \frac{1}{2}] + [\frac{1}{2} \div \frac{1}{2}] = 2$
7. $(2 - 2) - (2 - 2) = 0$
8. $[\frac{1}{2} \div \frac{1}{2}] + [\frac{1}{2} - \frac{1}{2}] = 1$
9. $[\frac{1}{2} \cdot \frac{1}{2}] + [\frac{1}{2} \cdot \frac{1}{2}] = \frac{1}{2}$
10. $[2 \div \frac{1}{2}] - [2 \cdot \frac{1}{2}] = 3$
11. $[\frac{1}{2} \cdot 2] - [\frac{1}{2} \div 2] = \frac{3}{4}$

Practice sheet 3

1. $6 \div 1$, $1 \div \frac{1}{6}$
2. $1 \div 8$
3. $6 \cdot \frac{1}{7}$, $6 \div 7$
4. $10 \div 9$
5. $4 \div 5$
6. $9\overline{)8}$
7. $2\overline{)3}$, $3 \div 2$, $3 \cdot \frac{1}{2}$

Practice sheet 4

Answers will vary.

Practice sheet 5

1. $>$, $\neq 1$
2. $=$, $= 1$
3. $>$, $\neq 1$
4. $>$, $\neq 1$
5. $>$, $\neq 1$
6. $=$, $= 1$
7. $=$, $= 1$
8. $<$, $\neq 1$

Practice sheet 6

1. < 1
2. $= 1$
3. > 1
4. < 1
5. $= 1$
6. > 1
7. $= 1$
8. < 1
9. > 1
10. < 1

Practice sheet 7

1. $= 17$
2. $= 17$
3. $= 17$
4. $= 17$
5. < 17
6. > 17
7. < 17
8. < 17
9. < 17
10. < 17

Practice sheet 8

1. $\frac{7}{3}$
2. $\frac{5}{9}$
3. $\frac{21}{2}$
4. $\frac{42}{19}$
5. $\frac{21}{16}$

Practice sheet 9

Answers will vary. For example,

1. 16, $14 + 2 = 16$
2. 33, $27 + 2 = 29$
3. 12, $4 + 3 + 2 = 9$
4. 36, $20 + 63 = 8 + 75$
5. 16, $16 = 14 + 3 - 1$
6. 30, $15 + 40 = 67 - 12$
7. 12, $42 = 30 - 60 + 72$
8. 21, $19 + 42 = 49 + 12$

Practice sheet 10

Answers will vary. For example,

1. 10, $4 + 5 = 9$
2. 36, $18 + 10 = 28$
3. 21, $7 - 6 = 9 - 8$

4. 8, $8 + 5 = 3 + 6 + 4$

5. 18, $12 + 10 + 21 = 72 - 21 - 8$

6. $2 \cdot 3 \cdot 3$, $15 + 14 = 29$

7. $2 \cdot 2 \cdot 3 \cdot 5$, $15 - 8 = 10 - 3$

8. $2 \cdot 2 \cdot 2 \cdot 3 \cdot 3 \cdot 5$, $84 = 45 + 40 - 55 + 54$

Practice sheet 11

1. a) multiply each fraction by 1 to get a common denominator

b) rewrite each term with the l.c.d.

c) combine the terms

d) factor the numerator and the denominator

e) rewrite without the 1's

2. a) multiply each term by the l.c.d.

b) rewrite each term

c) simplify each term

d) perform the indicated operations

3. In question 1 an expression is being simplified. In question 2 an equation is being cleared of fractions.

Practice sheet 12

1. a) no, 1 or any name for 1

b) no, 1 or any name for 1

c) yes

d) yes

e) no, 1 or any name for 1

2. a) no; incorrect for adding fractions, does not preserve value of fractions

b) yes; incorrect for adding, no common denominator

c) yes; incorrect for adding, no common denominator

d) yes; correct for adding, common denominator

Practice sheet 13

1. a) Not equal; $\frac{5}{7} \neq 5$, $3 \neq 21$.
(Remind students that they are not eliminating fractions in an equation here.)

b) Equal; $\frac{28}{4}$ is another name for 7

c) Equal; using the multiplicative property of 1 and another name for 6

d) Not equal; $\frac{3}{2} \neq 3 \cdot 3$, $\frac{5}{3} \neq 2 \cdot 5$.
(It is not appropriate to "cross multiply" here.)

e) Equal; using another name for 1

and the multiplicative property of 1

f) Not equal; incorrect method for adding fractions; need a common denominator

g) Equal; using another name for 1 and the multiplicative property of 1

2. a) no; $\frac{\square}{3} \neq \square$, $4 \neq 12$.

b) yes; using another name for 1 and the multiplicative property of 1

c) yes; using another name for 1 and the multiplicative property of 1

d) yes; using rule for subtracting fractions

e) no; value of fractions not preserved (haven't multiplied by 1)

Practice sheet 14

1. no

2. yes

3. yes

4. yes

5. yes

6. yes

7. no

8. no

9. yes

10. yes

11. no

12. no

13. yes

Practice sheet 15

1. a) yes, 21

b) no

c) no

d) yes, 0

e) yes, 42

f) no

2. D

A is incorrect because each term has not been multiplied by 8. ($\frac{3}{4}$ was multiplied by $\frac{2}{2}$ and 2 was expressed as $\frac{16}{8}$.)

B is incorrect for two reasons; $\frac{3}{4} \cdot 8 \neq 3$ and the right-hand member of the equation was not multiplied by 8.

C is incorrect because the right-hand member of the equation was not multiplied by 8.

Fractions in Expressions and Equations

Answers to Follow-up Assessment

1. a) no   **e)** yes   **c)** no   **e)** no   **d)** no   **c)** yes
  b) yes   **f)** no   **d)** no   **f)** yes   **21.** C   **d)** yes
  c) yes   **5. a)** yes   **e)** yes   **g)** yes   **22. a)** yes   **e)** no
  d) no   **b)** no   **10. a)** no   **15. a)** yes   **b)** no   **26. a)** no
  e) no   **c)** yes   **b)** no   **b)** yes   **c)** no   **b)** no
  f) no   **d)** yes   **c)** yes   **c)** no   **d)** yes   **c)** yes
2. a) yes   **e)** no   **d)** yes   **d)** yes   **23. a)** no   **d)** no
  b) yes   **6. a)** yes   **e)** yes   **e)** yes   **b)** yes   **e)** yes
  c) no   **b)** yes   **11.** 0   **f)** no   **c)** yes   **27. a)** no
  d) yes   **c)** no   **12.** Any   **16.** Answers   **d)** no   **b)** no
  e) yes   **d)** yes   nonzero   will vary.   **e)** yes   **c)** no
  f) yes   **e)** no   number   **17.** B   **f)** yes   **d)** no
3. a) no   **7.** Answers   **13. a)** yes   **18.** A   **24. a)** yes   **e)** no
  b) yes   will vary.   **b)** no   **19. a)** no   **b)** yes   **f)** yes
  c) no   **8. a)** no   **c)** yes   **b)** yes   **c)** yes   **g)** yes
  d) yes   **b)** yes   **d)** no   **c)** yes   **d)** yes   **28.** Answer
  e) yes   **c)** yes   **e)** no   **d)** no   **e)** yes   will vary.
4. a) yes   **d)** yes   **14. a)** yes   **e)** no   **f)** no   For
  b) yes   **e)** no   **b)** no   **20. a)** yes   **g)** no   example,
  c) yes   **9. a)** yes   **c)** no   **b)** yes   **25. a)** no   **a)** $18+7=25$
  d) yes   **b)** yes   **d)** yes   **c)** no   **b)** no   **b)** 12

ALGE|B|R|I|D|G|E ™

Meaning

of Negative

Numbers

Meaning of Negative Numbers

Introduction

Students have little difficulty understanding the relationship between three apples and the number 3, but developing a similar relationship for the number $^-3$ is often perplexing for them. This unit helps students to develop models of negative numbers so that they understand what a negative number represents. Such an understanding can help students overcome or avoid difficulties in working with negative numbers when they are introduced to the rules that govern the mathematical manipulation of negative numbers.

In the context of realistic situations, this unit introduces models and uses a number line in different ways to help students to

– develop the idea of opposites;

– identify ways to represent signed numbers;

– recognize the order of signed numbers;

– determine the relative value of signed numbers;

– understand implied relationships between signed numbers.

Cross-References to
Pre-Algebra and Algebra
Textbook Topics

The concepts addressed in this unit are often not adequately covered in textbooks. Most textbooks focus on operations with signed numbers and do not provide much help in developing a sense of what a negative number means.

This unit may be helpful when studying the following topics that appear in many pre-algebra and algebra textbooks.

integers **integers and equations**

List of Terms

The following is a list of words and phrases that are frequently used when discussing signed numbers. The teacher may want to add other words or phrases to this list.

absolute value	**neutral**	**signed number**
Celsius	**number line**	**temperature**
degree	**opposite**	**thermometer**
Fahrenheit	**positive number**	**zero**
negative number	**scale**	

Classification by Concept

Target Concepts and Abilities	Questions in Instructional Assessment	Questions in Follow-up Assessment	Practice Sheets
Develop the idea of opposites	1, 2, 3, 18	1, 2, 3, 15	
Identify ways to represent signed numbers	4, 6, 9, 10	4, 6	1, 2, 3, 4
Recognize the order of signed numbers	6, 16, 17, 18	6, 15	3, 4, 6, 7
Determine the relative values of signed numbers	3, 4, 5, 7, 8, 9, 10, 11, 13, 14, 15, 16, 17, 18	3, 4, 5, 7, 8, 9, 11, 12, 13, 14, 15	2, 3, 4, 5, 8, 9, 10, 11, 12, 13
Understand implied relationships between signed numbers	7, 8, 9, 10, 11, 12, 14, 15	8, 9, 10, 11, 12	8, 9

Classification by Question

Target Concepts and Abilities	Develop the idea of opposites	Identify ways to represent signed numbers	Recognize the order of signed numbers	Determine the relative value of signed numbers	Understand implied relationships between signed numbers
Question Number or Practice Sheet Number 1	IA FA	PS			
2	IA FA	PS		PS	
3	IA FA	PS	PS	IA FA PS	
4		IA FA PS	PS	IA FA PS	
5				IA FA PS	
6		IA FA	IA FA PS		
7			PS	IA FA	IA
8				IA FA PS	IA FA PS
9		IA		IA FA PS	IA FA PS
10		IA		IA PS	IA FA
11				IA FA PS	IA FA
12				FA PS	IA FA
13				IA FA PS	
14				IA FA	IA
15	FA		FA	IA FA	IA
16			IA	IA	
17			IA	IA	
18	IA		IA	IA	

IA - Questions in Instructional Assessment
FA - Questions in Follow-up Assessment
PS - Practice Sheets

Instructional Assessment

The questions in both the Instructional and Follow-up Assessment Instruments for this unit have different formats appropriate to the nature of the concepts addressed in the questions.

It is necessary to introduce students to these formats since they may have limited experience with different types of responses.

If there is not sufficient time to review examples of the different question types, review the directions below with the students before they work the *Algebridge*™ exercises.

Directions

There are three types of questions in this exercise.

One type asks you to choose the answer to the problem and mark the corresponding letter, A, B, C, or D, on your answer sheet.

The second type gives you several choices related to a specific question. You must answer **yes** or **no** for each of these choices.

The third type asks you to write your own answer to the problem.

Focus of Questions 1–2

Questions 1 and 2 introduce the concept of opposites. This concept is the underlying basis of the idea of negative numbers: negative is the opposite of positive.

Questions 1-2

1. Here is a list of opposites. For example, the opposite of *Hot* is *Cold*.
   Fill in the missing opposites.

	Word	**Opposite**
Example	Hot	Cold
a)	Below	*Above*
b)	*10 hours before*	10 hours after
c)	West	*East*
d)	Lose	*Win*
e)	*Sell*	Buy
f)	Ahead 5 paces	*Behind 5 paces*
g)	Earn money	*Spend Money*
h)	*Clockwise*	Counterclockwise
i)	Gain 3 pounds	*Lose 3 pounds*
j)	Advance	*Retreat/Go Back*
k)	*1 mile south*	1 mile north
l)	*Deposit $5*	Withdraw $5
m)	Shrink	*Stretch/Expand*

2. Here is a list of words often used in mathematics. Fill in the missing opposites.

	Word	**Opposite**
a)	Plus	*Minus*
b)	*Addition*	Subtraction
c)	*Positive*	Negative
d)	3 more than	*3 less than*
e)	*Positive 8*	Negative 8
f)	*Right*	Left
g)	Increase	*Decrease*
h)	Above	*Below*

Analysis of Questions 1–2

Question 1 encourages students to draw from their own experiences to identify familiar ideas of opposites. Some of the students' answers may differ from those given in the teacher's edition. If students can provide valid rationales for their answers, they should be given credit.

Question 1 is meant to lead into question 2, which deals solely with mathematical ideas and terminology, in the hope that students may be able to transfer the idea of opposites in their own experience to the more abstract ideas in mathematics.

Guiding Class Discussion

The words used in questions 1 and 2 often appear in word problems. The teacher could ask the students to indicate which of the words in each pair would be assigned a negative value if it appeared in the context of a word problem. In some cases, such as "East/West," the student should see that either of the two words could be assigned the negative value, but it is often more natural to assign a negative value to one member of a pair than to the other. It is possible for a word to be considered positive in one context, but negative in another. For example, depositing $5 in an account can be considered positive relative to the account, but negative relative to the available cash-on-hand.

To emphasize the concept of the opposite of a number and the symbolism for positive and negative numbers, the teacher could write several signed numbers on the chalkboard. The students could be asked to write the opposite of each number. The list might look like this:

Number	Opposite of the Number
$^+8$	_____
$^-7$	_____
$^+5$	_____
$^-2$	_____
0	_____
$^+1$	_____

The students could plot these numbers and their opposites on a number line. A discussion might follow about the fact that 0 is always in the middle between any number and its opposite. Some textbooks say that 0 does not have an opposite and some say that 0 is its own opposite.

The practice sheets and suggested activities in the Follow-up Instruction section under the concept **Develop the idea of opposites** may be helpful for those students having difficulty or needing more practice.

Focus of Question 3

This question expands the idea of opposites to include the concept of a neutral point. Again, students are encouraged to draw from their own experiences in order to see that the ideas of positive, neutral, and negative do not belong exclusively to the world of mathematics.

Question 3

3. In the table below, the words in each row refer to three ways to look at a situation. Fill in the missing parts. Part a is done for you.

	Situation	Negative	Neutral	Positive
a)	Time	Yesterday	Today	Tomorrow
b)	Business	*Lose*	Break even	Gain
c)	Sports	Lose	Draw/Tie	*Win*
d)	Board games	Go back	*Stay put*	Go forward
e)	Elevation	Below sea level	Sea level	*Above sea level*
f)	Time	*Before*	Now	After
g)	Position	Left	*Middle*	Right
h)	Mathematics	Negative number	*Zero*	Positive number

Guiding Class Discussion

An important point to discuss with the students at this time is the idea that "neutral" (or "zero") can represent a point of reference and does not necessarily represent the absence of a quantity. On a board game, for example, to "stay put" does not necessarily mean that one is at the starting point of the game; it can mean that the player's position is halfway around the game board, but he or she did not move during a particular turn in the game.

Another idea involved in this question is that there are often natural tendencies for the definition of "positive."

To explore further the ideas of "neutral" and "point of reference," the teacher might suggest a temperature setting in which "negative, neutral, positive" would be "too cold, just right, too hot." The discussion of the example should include the idea that "just right," even though it represents the neutral temperature point, is not necessarily 0°. Rather, if one considers room temperature, "just right" would be about 22°C (72°F). Temperatures less than 22°C are "too cold" (negative) and those above 22°C are "too hot" (positive). In this case 22°C is a point of reference that can be useful in judging how hot or cold a particular temperature is. Showing the thermometer at the left to students might help them to see that 0 on a temperature scale is a **relative point** since 0 is not at the same position for both the Celsius and Fahrenheit scales. This may confuse students, so it is important to reinforce the idea that 0 is also used as a **reference point**. In the case of numbers, numbers less than 0 are negative and those greater than 0 are positive. Contrast this with the temperature example in which 22° is the reference point, where a temperature that is "too cold" or "negative" may have a positive value (*e.g.*, $^+4°$C). The development of this idea will help students later when they are introduced to the number line, where they must distinguish between the value of a position on the number line and the value and direction of a movement along the number line. It may be helpful to have students offer examples of other situations in which a negative, neutral, and positive aspect can be identified.

Asking the students to assign a negative or positive number to each of the terms in the "negative" and "positive" columns of question 3 and having them offer examples of situations in which positive direction is relative may help students to see why it is useful to assign signs to numbers and thus why negative numbers can be useful.

As in questions 1 and 2, the teacher might discuss the idea that there are often natural tendencies for the definition of "positive," right and above sea level, for example. Students should be made aware, however, that there are some situations in which positive direction is relative. For example, in a coin toss, heads or tails could be positive, depending on which side of the coin is chosen.

Focus of Question 4

Question 4 introduces the thermometer as a real-world model of a number line. Note that this unit adopts the convention of a raised "+" and "−" for positive and negative numbers so that students do not confuse the signs for positive and negative with those for addition and subtraction.

Question 4

4. Thermometer 1 reads 5 degrees above zero. A temperature reading of 5 degrees above zero can be written as $^+5$.

   Thermometer 2 reads 6 degrees below zero. A temperature reading of 6 degrees below zero can be written as $^-6$.

   $^+5$ and $^-6$ are signed numbers.

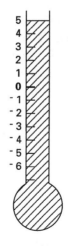

Thermometer 1

Thermometer 2

The table below shows temperature readings. Fill in their corresponding signed numbers.

	Temperature Reading	Signed Number
a)	8 degrees below zero	$^-8$
b)	9 degrees above zero	$^+9$
c)	7 degrees	$^+7$
d)	0 degrees	0
e)	15 degrees above zero	$^+15$
f)	8 degrees above 5 degrees	$^+13$
g)	8 degrees below 5 degrees	$^-3$

Guiding Class Discussion

This question presents another opportunity for discussing the idea of 0 being a relative point. Although temperatures are often reported as being a number of degrees above or below 0, it is equally valid to say that a temperature is a number of degrees above or below some other reference point, such as the freezing point of water (32°F). Questions 4f and 4g introduce this idea, using 5 degrees as the reference point. Discussion could focus on the ease of using 0 degrees as the reference point when reporting temperatures.

The teacher should discuss with the students that the positive sign is often assumed; that is $^+7$ is often written as 7. In question 4c, the students should be able to recognize that "7 degrees" indicates "7 degrees above zero." The teacher should help them to see that a similar convention is used in mathematics, namely that "7" indicates "$^+7$." In other words, an unsigned number other than 0 is taken as positive.

The practice sheets and suggested activities in the Follow-up Instruction section under the concepts **Identify ways to represent signed numbers** and **Determine the relative values of signed numbers** may be helpful for those students having difficulty or needing more practice.

In particular, **practice sheets 1 and 2** use a thermometer as a model for representing signed numbers. For practice sheet 1, students are given a temperature indicated by a filled-in thermometer. The student writes the corresponding temperature reading and a signed number. For practice sheet 2, the temperature reading is given and the student writes the corresponding signed number and fills in the thermometer to indicate the temperature. **Practice sheets 3 and 4** ask students to do the same tasks using a semi-circular dial as the model.

Focus of Question 5   Question 5 introduces the idea of relative value.

_____

Question 5

5. For each of the following pairs of temperature readings, circle the *colder* temperature.

  a) ⟨⁺4⟩ ⁺5
  b) ⁺5 ⟨⁻3⟩
  c) ⁺4 ⟨⁻4⟩
  d) ⟨⁻4⟩ 0
  e) ⁻3 ⟨⁻4⟩
  f) ⟨⁻5⟩ ⁻2
  g) ⟨0⟩ ⁺4

_____

Analysis of Question 5

Students are probably familiar with the idea that a lower thermometer reading indicates a colder temperature. This question gives students an opportunity to transfer ideas they have developed about relative value from their own experience to signed numbers on a number line. If they can understand that 4 degrees below zero is colder than 3 degrees below zero, the teacher may be able to convince them that $^-4 < {}^-3$, and, similarly, that $^-4 < 0$.

Guiding Class Discussion

The practice sheets and suggested activities in the Follow-up Instruction section under the concept **Determine the relative values of signed numbers** may be helpful for those students having difficulty or needing more practice.

In particular, **practice sheet 5** presents pairs of thermometers allowing the student to represent and compare the temperatures given in each part of question 5.

Focus of Question 6

In question 6, the students must create their own vertical number line. This exercise helps them to see that, although zero can be placed anywhere on the thermometer, its position should be chosen in such a way that all the given numbers can fit on the thermometer scale.

Question 6

6. Label the thermometer with the following temperatures.

    a)   +4
    b)   0
    c)   −4
    d)   +6
    e)   +3
    f)   −2

Guiding Class Discussion

The teacher should discuss with the students what happens once one number is written on the thermometer and then what happens once a second number is written. They need to see that, once a scale is established, the positions of the other numbers are already determined.

Students who have trouble seeing that the position of zero on a scale can be arbitrarily chosen may put 0 at the bottom of the thermometer. Bringing a real thermometer into the classroom may help convince students that zero does not have to be placed at the bottom of the scale.

If students order the numbers correctly but put the negative numbers above zero and the positive numbers below zero, it would be helpful to discuss with them increasing and decreasing order and how to determine which type of order is appropriate in a given situation.

The practice sheets and suggested activities in the Follow-up Instruction section under the concepts **Identify ways to represent signed numbers** and **Recognize the order of signed numbers** may be helpful for those students having difficulty or needing more practice.

In particular, **practice sheets 6 and 7** give more practice in labeling thermometers and writing signed numbers to represent temperatures.

Focus of Questions 7–8

Question 7 integrates several important ideas associated with the number line: value of a position on the number line (the given temperatures), direction of a movement along the number line (whether the temperature goes up or down), and magnitude of a movement along the number line (amount of change in the temperature).

In question 8, students are also shown how the direction of a movement along the number line (in this case, the temperature scale) determines the sign that is assigned to the magnitude of that change.

Questions 7–8

7. For each part of this question, temperature readings at 12 noon on two different days are given for a certain city. Mark the temperatures on the thermometer scales. Indicate whether the noontime temperature went up or down from day 1 to day 2 by circling **up** or **down**. Then write by how many degrees the temperature changed. Parts a and b are done for you.

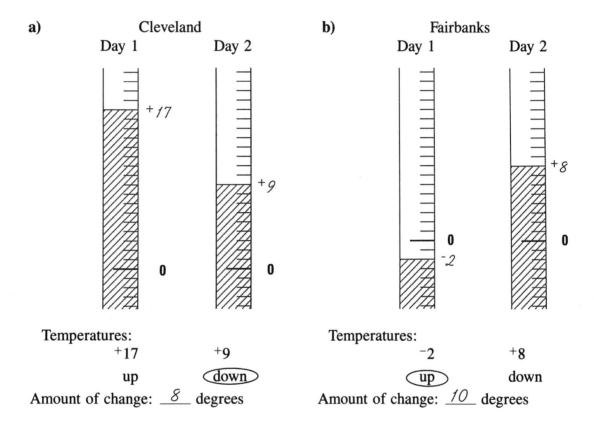

a)   Cleveland   b)   Fairbanks

Temperatures: Temperatures:
    +17   +9   −2   +8
    up   (down)   (up)   down
Amount of change: __8__ degrees   Amount of change: __10__ degrees

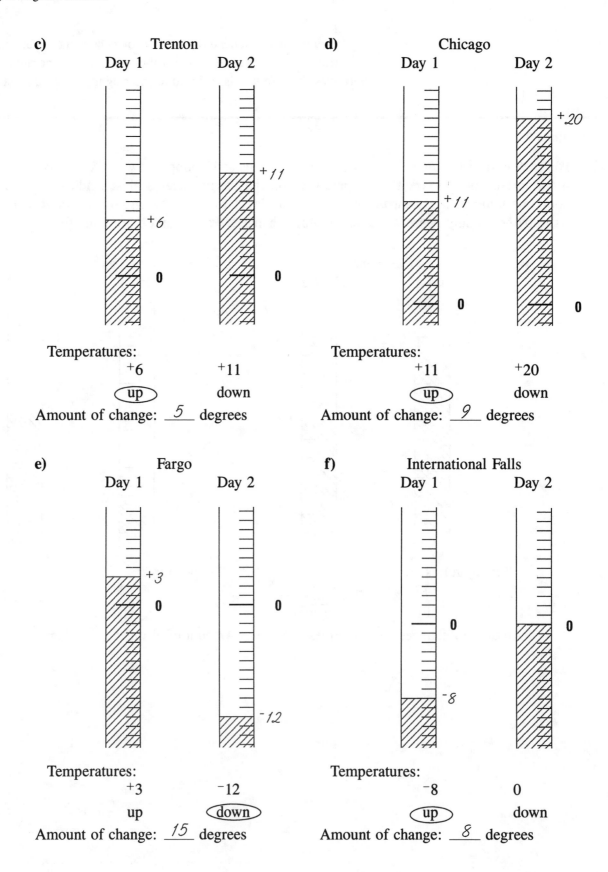

c)   Trenton

Day 1   Day 2

⁺6

⁺11

0   0

Temperatures:

⁺6   ⁺11

(up)   down

Amount of change: __5__ degrees

d)   Chicago

Day 1   Day 2

⁺20

⁺11

0   0

Temperatures:

⁺11   ⁺20

(up)   down

Amount of change: __9__ degrees

e)   Fargo

Day 1   Day 2

⁺3

0   0

⁻12

Temperatures:

⁺3   ⁻12

up   (down)

Amount of change: __15__ degrees

f)   International Falls

Day 1   Day 2

0   0

⁻8

Temperatures:

⁻8   0

(up)   down

Amount of change: __8__ degrees

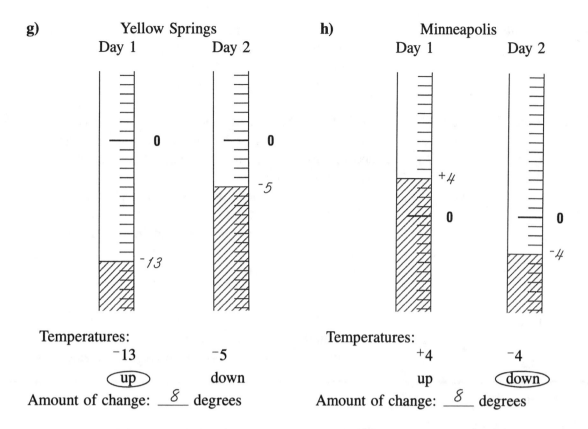

g)   Yellow Springs

Day 1   Day 2

Temperatures:

-13   ⁻5

⟨up⟩   down

Amount of change: _8_ degrees

h)   Minneapolis

Day 1   Day 2

Temperatures:

⁺4   ⁻4

up   ⟨down⟩

Amount of change: _8_ degrees

---

8. The thermometers in question 7 show that in Cleveland the temperature went down 8 degrees. This change in temperature can be written as ⁻8. In Fairbanks, the temperature went up 10 degrees. This change in temperature can be written as ⁺10.

For each city write a signed number to show the change in temperature at noon from day 1 to day 2. The thermometers in each part of question 7 may help you to do this.

	City	Day 1	Day 2	Change
a)	Cleveland	⁺17	⁺9	⁻8
b)	Fairbanks	⁻2	⁺8	⁺10
c)	Trenton	⁺6	⁺11	⁺5
d)	Chicago	⁺11	⁺20	⁺9
e)	Fargo	⁺3	⁻12	⁻15
f)	International Falls	⁻8	0	⁺8
g)	Yellow Springs	⁻13	⁻5	⁺8
h)	Minneapolis	⁺4	⁻4	⁻8

Analysis of Questions 7–8

Question 7 leads the students through each of the ideas stated above and gives students a model to follow when they are asked to put these concepts together in question 8. Note that the amount of change in question 7 is the absolute value of the change.

Guiding Class Discussion

Students may be confused about the sign associated with the change in temperature. It may be helpful to remind them of earlier discussions about positive, neutral (or zero), and negative and their association with increase, no change, and decrease.

The practice sheets and suggested activities in the Follow-up Instruction section under the concepts **Determine the relative values of signed numbers** and **Understand implied relationships between signed numbers** may be helpful for those students having difficulty or needing more practice.

Focus of Question 9

Question 9 presents to the students a problem that they can solve with or without the use of the thermometer model. This question is meant to help students see the processes involved in analyzing such a problem.

Question 9

9. On March 9 at 10 a.m., the temperature in Kansas City was 39 degrees. The temperature rose 3 degrees by noon and then fell 5 degrees by 5 p.m.

   a) Between 10 a.m. and 5 p.m. did the temperature fall more degrees than it rose?

   Answer: *Yes*

   b) At 5 p.m., was the temperature above or below 39 degrees?

   Answer: *Below*

   c) How much above or below 39 degrees was the temperature at 5 p.m.?

   Answer: *2 degrees below 39*

   d) What was the temperature at 5 p.m.?

   Answer: *37 degrees*

Analysis of Question 9

Question 9d is what textbooks usually ask. Questions 9a, 9b, and 9c help the students analyze conditions of the problem so that they can more easily judge whether their response to question 9d is correct.

Guiding Class Discussion

The practice sheets and suggested activities in the Follow-up Instruction section under the concepts **Identify ways to represent signed numbers**, **Determine the relative values of signed numbers**, and **Understand implied relationships between signed numbers** may be helpful for those students having difficulty or needing more practice.

Focus of Question 10
Question 10 follows the same pattern as question 9. In this question, a negative starting point is used.

Question 10

10. On January 26 at 10 a.m., the temperature in Milwaukee was ⁻14 degrees. The temperature rose 11 degrees by noon and then fell 7 degrees by 5 p.m.

    a) Between 10 a.m. and 5 p.m. did the temperature rise more degrees than it fell?

       Answer: *Yes*

    b) At 5 p.m., was the temperature above or below ⁻14?

       Answer: *Above*

    c) How much above or below ⁻14 was the temperature at 5 p.m.?

       Answer: *4 degrees above ⁻14*

    d) What was the temperature at 5 p.m.?

       Answer: *⁻10 degrees*

Guiding Class Discussion
Students who find question 10 challenging should be encouraged to refer to question 9 and to use it as a model for solving this question.

The practice sheets and suggested activities in the Follow-up Instruction section under the concepts **Identify ways to represent signed numbers**, **Determine the relative values of signed numbers**, and **Understand implied relationships between signed numbers** may be helpful for those students having difficulty or needing more practice.

In particular, **practice sheets 1 and 2** could be used if students are confused about reading temperature scales. Also, students could use the thermometers of **practice sheet 5** to indicate the problem situation in question 9 or 10.

Focus of Questions 11–12
Questions 11 and 12 extend some of the ideas previously introduced in this unit. The students are now asked to evaluate the magnitude of change associated with signed numbers.

Questions 11–12

11.   **a)** On April 11 in Milwaukee, between 10 a.m. and 12 noon, the temperature changed ⁻6 degrees. Did the temperature go up or down during this period?

       Answer: *Down*

b) Between noon and 5 p.m. on that day, the change in temperature was $^+2$ degrees. Did the temperature go up or down during this period?

Answer: *Up*

c) Was the temperature at 5 p.m. above or below the temperature at 10 a.m.?

Answer: *Below*

d) By how many degrees?

Answer: *4*

12. The table below shows the temperature changes in Milwaukee on several other days. For each day, indicate the apparent change in temperature between 10 a.m. and 5 p.m.

	Day	Change in Temperature from 10 a.m. to Noon	Change in Temperature from Noon to 5 p.m.	Change in Temperature from 10 a.m. to 5 p.m.
a)	April 12	$^+3$	$^+6$	$^+9$
b)	April 13	$^+7$	$^-2$	$^+5$
c)	April 14	$^+4$	$^-5$	$^-1$
d)	April 15	0	$^+4$	$^+4$
e)	April 16	0	$^-6$	$^-6$
f)	April 17	$^+5$	0	$^+5$
g)	April 18	$^-3$	0	$^-3$
h)	April 19	$^-6$	$^+4$	$^-2$
i)	April 20	$^-2$	$^-5$	$^-7$

Guiding Class Discussion

It is critical at this point to discuss with the students the distinction between $^-6$ as a temperature and $^-6$ as a *change* in temperature.

If students have difficulty with question 12, they can choose a temperature as a starting point, calculate the changes in temperature that are given in the table, and then compare the final temperature with the starting temperature. Students should be able to see that they can choose any starting temperature and still obtain the same answer.

The practice sheets and suggested activities in the Follow-up Instruction section under the concepts **Identify ways to represent signed** numbers, **Determine the relative values of signed numbers**, and **Understand implied relationships between signed numbers** may be helpful for those students having difficulty or needing more practice.

Focus of Question 13

Question 13 introduces the idea of absolute value, which surfaces early in algebra.

Question 13

13. For each of the following pairs of temperature readings, circle the one that is farther from zero.

a) (⁺6) ⁺5
b) (⁻4) ⁺2
c) (⁻3) ⁻1
d) ⁻4 (⁺5)
e) 0 (⁻3)
f) (⁺6) ⁺4
g) (⁺6) ⁻4
h) (⁻6) ⁻4

Analysis of Question 13

Note that the absolute values of the numbers in the last three pairs of numbers are the same, namely 6 and 4, although the signs of the numbers are different. Another question to ask the students is which number they would circle if the numbers were ⁺8 and ⁻8. They should be able to see that ⁺8 and ⁻8 are equally distant from zero.

Guiding Class Discussion

The practice sheets and suggested activities in the Follow-up Instruction section under the concept **Determine the relative value of signed numbers** may be helpful for those students having difficulty or needing more practice.

In particular, students could use the thermometers on **practice sheet 5** to compare the temperatures given in each part of question 13.

Focus of Questions 14–15

Questions 14 and 15 focus on the relative values of signed numbers.

Questions 14–15

Questions 14–15 refer to the following information.

The rules for a card game are described below. In a deck of cards, half of the cards have a [+] sign and the others have a [−] sign.

A [+] card is worth ⁺1.
A [−] card is worth ⁻1.

Each player chooses five cards from the deck.
The player whose score is the greater number wins.

For example, Bianca and Jason play a game. The cards they chose and their scores are shown below.

						Score
Bianca	[+]	[−]	[+]	[+]	[−]	+1
Jason	[−]	[−]	[+]	[−]	[−]	‾3

Bianca won the game because her score, +1, is greater than Jason's score, ‾3.

14. Several other students play the game. The cards they chose are shown below. Fill in the scores and write the name of the winner.

								Score	**Who Won?**
a)	Latreese	[+]	[+]	[+]	[+]	[−]		+3	*Latreese*
	Calvin	[−]	[−]	[−]	[−]	[+]		‾3	
b)	Nancy	[−]	[−]	[−]	[−]	[−]		‾5	*Jeremy*
	Jeremy	[+]	[+]	[+]	[+]	[+]		+5	
c)	Dana	[−]	[+]	[−]	[+]	[−]		‾1	*Yolanda*
	Yolanda	[+]	[+]	[+]	[+]	[−]		+3	

15. In another game with these cards, the players can choose how many cards they will get. Each player can get up to ten cards. The player whose score is the greater number wins.

Several students play this game. The cards they chose are shown below. Fill in the scores and write the name of the winner.

													Score	**Who Won?**
a)	Denise	[+]	[+]	[−]	[+]	[+]	[−]						+2	*Brian*
	Brian	[+]	[−]	[+]	[+]	[+]							+3	
b)	Greg	[−]	[−]	[+]	[−]	[−]	[−]						‾4	*Neither*
	Amy	[−]	[−]	[−]	[+]	[−]	[+]	[−]	[−]				‾4	
c)	Eloiza	[−]	[−]	[−]									‾3	*Eloiza*
	Gabe	[−]	[−]	[−]	[+]	[+]	[−]	[+]	[−]	[−]	[−]		‾4	
d)	Brent	[+]	[+]	[−]	[−]								0	*Neither*
	Chris	[−]	[+]	[+]	[+]	[−]	[−]						0	

Analysis of Questions 14–15

Although these questions do introduce the idea of implied addition of signed numbers, it is important for students to understand the meaning of signed numbers before they are introduced to operations with these numbers. These questions also serve to give the students a model for signed numbers that will be helpful to use as their work with signed numbers becomes more complex.

To avoid discussion of the rules for addition of signed numbers, it may be helpful to remind students that since a "+" has a value of $^+1$ and a "−" has a value of $^-1$, a [+] and a [−] combine to give zero. After eliminating "zero pairs," the student will need only to count the remaining negatives or positives.

The practice sheets and suggested activities in the Follow-up Instruction section under the concept **Determine the relative values of signed numbers** and **Understand implied relationships between signed numbers** may be helpful for those students having difficulty or needing more practice.

Practice Sheet 8 is a master for making a set of game pieces so that students can play the games described in questions 14 and 15. **Practice Sheet 9** can be used as a score sheet for games played with the [+] and [−] cards.

Focus of Question 16

Question 16 introduces a number line.

Question 16

16. The following questions refer to the number line to the right.

a) Which arrow points to $^-2$? _S_

b) Which arrow points to 0? _H_

c) Which arrow points to $^-3\frac{1}{2}$? _W_

d) Which arrow points to $^+\frac{1}{2}$? _D_

e) Of the arrows to the right, which one points to the greatest number? _E_

Analysis of Question 16

Since the model of the thermometer was used throughout this unit, students may have an easier time transferring what they have learned about signed numbers on the thermometer to a vertical number line that resembles this model. Question 16c presents the idea that negative numbers do not always have to be integers.

Guiding Class Discussion

The practice sheets and suggested activities in the Follow-up Instruction section under the concept **Determine the relative values of signed numbers** may be helpful for those students having difficulty or needing more practice.

For example, **practice sheets 10 and 11** could be used to give students more practice with fractions and mixed numbers on a scale.

Focus of Question 17

Question 17 introduces students to a traditional number line.

Question 17

17.

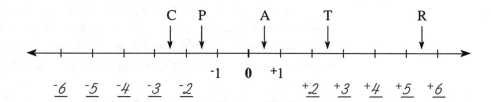

a) On the number line above, fill in the blanks with the missing signed numbers.

b) Which arrow points to +2.5?   *T*

c) Which arrow points to ⁻1.5?   *P*

d) Which arrow points to the number that is closest to zero?   *A*

e) Of the arrows above, which points to the least number?   *C*

Guiding Class Discussion

If students have difficulty with this question, the teacher should show them how this number line is like the vertical number line (or the thermometer) except that it is horizontal. A discussion which brings out the idea that once the neutral point (zero) and a unit length are specified, all other points on the number line are uniquely determined would be helpful. Question 17c, like question 16c, shows the students that numbers that are not integers can be negative.

The practice sheets and suggested activities in the Follow-up Instruction section under the concept **Determine the relative values of signed numbers** may be helpful for those students having difficulty or needing more practice.

For example, **practice sheets 12 and 13** could be used to provide more practice with decimal numbers on a scale.

Focus of Question 18

Question 18 expresses in words many of the basic concepts introduced in the unit. Each of these statements expresses a generalization that students should be able to make if they have understood the material in the unit.

Question 18

18.

Consider each of the following. Is the statement true about the number line above?

yes   no   **a)** Any number to the right of zero is greater than any number to the left of zero.

yes   no   **b)** Any number is greater than a number to its left.

yes   no   **c)** The opposite of any number to the right of zero is always to the left of zero.

yes   no   **d)** The farther a number is from zero, the greater is its value.

yes   no   **e)** Zero is neither positive nor negative.

yes   no   **f)** Any number to the left of zero is always greater than zero.

yes   no   **g)** The opposite of any number to the left of zero is always to the right of zero.

Guiding Class Discussion

If students have difficulty defending their answers to these questions, using a number line or examples from other questions in the unit to illustrate the concept would be helpful.

For example, for 18a the students could draw a horizontal number line and fill in the numbers. The students can create a table with two columns: "Numbers to the right of zero" and "Numbers to the left of zero." Then they can choose, at random, numbers from the right and left of zero on the number line and write them in the appropriate columns in the table. After the students have entered several positive and negative numbers in the table, they can use the fact that any positive number is greater than any negative number to conclude that the answer to question 18a is **yes**.

For 18d, as another example, the students could look at question 13 and note which numbers are circled. These should be the numbers farther from zero (that is, the numbers with the greater absolute value). The student should then determine if the circled number is always greater than the uncircled number. They should see that the number that is farther from zero is *not* always the number with the largest value. Therefore, the answer to question 18d is **no**.

The practice sheets and suggested activities in the Follow-up Instruction section under the concepts **Develop the idea of opposites**, **Recognize the order of signed numbers**, and **Determine the relative values of signed numbers** may be helpful for those students having difficulty or needing more practice.

Follow-up Instruction

After the Instructional Assessment has been used to pinpoint students' conceptual weaknesses, the teacher may wish to use the following suggested activities and practice sheets to help correct those misconceptions.

Develop the idea of opposites

Activity. Students could create their own word problems using some of the words in questions 1 and 2. For example, using the word pair "earn, spend," a student could write the following word problem: "In one week, Sue earned $10 babysitting and spent $4 on a record. How much money did she have at the end of the week?" Then the teacher could discuss with the students how to determine the assignment of a positive or negative value to the numbers in the word problem.

As a follow-up activity, students could add other mathematical words to the list in question 2 and keep the list in their notebooks for future reference and additions.

Questions 1–3 and 15 in the Follow-up Assessment also address this concept.

Identify ways to represent signed numbers

Practice sheets 1 and 2 use the thermometer and ask students to associate signed numbers with a temperature reading on the thermometer. **Practice sheets 3 and 4** parallel practice sheets 1 and 2 but use a different model. These sheets give the students another method of representing signed numbers. The dial used in sheets 3 and 4 is more "horizontal" than the thermometer and is consistent with the left-right to negative-positive correspondence on the number line.

Questions 4 and 6 in the Follow-up Assessment also address this concept.

Recognize the order of signed numbers

As in practice sheets 1 and 2, **practice sheets 6 and 7** ask students to write temperature readings in words and in symbols. In addition, they give students practice in creating their own "number line" by asking them to create the scale on the thermometer and the dial.

Questions 6 and 15 in the Follow-up Assessment also address this concept.

Determine the relative values of signed numbers

Activity. To give students practice in analyzing problems similar to that presented in question 9, the teacher could bring in a daily newspaper and have the students fill in tables like the one below. The tables could be put on the chalkboard or bulletin board. The class could divide into teams and each team could choose a city. Each team would follow the daily temperature changes for its city.

Date	Highest Temperature	Change in Highest Temperature from Previous Day
May 23	76	*
May 24	68	$^-8$
May 25	71	$^+3$

The teams in the class could use other tables in the newspaper, such as the information listed about stocks. Each team could choose a company to follow and record the change in price of the stock on a daily basis. A table similar to the one above could be posted for each company's stock.

Activity. The following activity may be a good way to introduce students to the operation of addition of signed numbers. If a deck of playing cards is available, the students could use them to play one of the games described in questions 14 and 15. Black cards could represent positive numbers (3 of clubs = $^+3$, 8 of spades = $^+8$) and red cards could represent negative numbers (2 of hearts = $^-2$, 10 of diamonds = $^-10$). Play may be less confusing if the face cards (jack, queen, and king) are removed. A number line could be drawn on the blackboard so that as each card is pulled by a player another student could point to the score of that player on the number line. For example, one player pulls the following cards from the deck:

5 of spades

3 of hearts

4 of diamonds

10 of clubs

6 of hearts

The student at the chalkboard would point successively to $^+5$, $^+2$, $^-2$, $^+8$, and $^+2$ for that player's game and the player's final score would be $^+2$. Through this exercise, students will be able to see how the score changes depending on the magnitude and direction of the movement on the number line. Also, students will be able to see that there are many ways in which to achieve a final score of $^-5$, for example.

Practice sheet 5 is a master that can be used in classroom activities for several questions in this unit. The teacher can write instructions that pertain to a specific activity on a photocopy of this practice sheet, and then make copies for the students to use.

For example, the teacher can use practice sheet 5 if students have difficulty with question 5. The students should write the first temperature of the pair next to a thermometer in column 1 and then fill in the thermometer. They should then do the same for the other temperature of the pair in column 2. The students should be able to see that the temperature associated with the colder day has the lesser value.

For additional practice, the teacher can give the students the following list of temperature readings and ask them to circle the temperature of the *warmer* day, using practice sheet 5 if they have difficulty.

Day 1	Day 2
$^+6$	$^+4$
$^+3$	$^-3$
$^-5$	$^+6$
$^-4$	$^-6$
$^+6$	$^-2$
$^-5$	$^-2$

Practice sheet 5 could also be used for students who have difficulty with question 13. The student should fill in the temperature readings in columns 1 and 2, respectively. Then they could mark off and measure the distance between 0 and top level of the "mercury" in each thermometer.

Practice sheets 10 and 11 can be used to reinforce the idea that signed numbers are not necessarily whole numbers. As in the given example, the students could draw scales on the dials of practice sheet 10 so that each arrow points to a different fraction. If the teacher chooses, practice sheet 10 could also be used for more practice with scales that use only whole numbers.

Practice sheets 12 and 13 are the same as practice sheets 10 and 11, but they involve using decimals rather than fractions. They can be used to reinforce the ideas in question 17. The students could draw scales on the dials of practice sheet 12 so that each arrow points to a decimal. If the teacher chooses, practice sheet 12 could also be used for more practice with scales that use only whole numbers.

Practice sheets 3 and 4 (discussed under the concept Identify ways to represent signed numbers) and **practice sheets 8–9** (discussed under the concept Understand implied relationships between signed numbers) also relate to the concept, Determine the relative values of signed numbers.

Questions 3–5, 7–9, and 11–15 in the Follow-up Assessment address this concept.

Understand implied relationships between signed numbers

Activity. The teacher could create a set of [+] and [−] cards using **practice sheet 8** and have the students play the games described for questions 14 and 15. (Chips or paper of two different colors could also be used, with positive assigned to one color and negative to the other.) **Practice sheet 9** can be used as a score sheet for each team of players.

One team of students could create new rules for a game using these cards. They could play the game in front of the class and fill in on the chalkboard a score sheet similar to practice sheet 9. The class would then have to guess the rules of the game. For example, one set of rules could be that each player gets five cards and the player whose score has the greatest absolute value would win. From the score sheet on the board, the class should be able to see a pattern in the scores to determine how the winner is chosen.

Questions 8–12 in the Follow-up Assessment also address this concept.

Bibliography of Resource Materials

Bartolini, Pietro. "Addition and Subtraction of Directed Numbers." *Mathematics Teacher 69* (March 1976): 34–35.

Chang, Lisa. "Multiple Methods of Teaching the Addition and Subtraction of Integers." *Arithmetic Teacher 33* (December 1985): 14–19.

Coon, Lewis H. "Number Line Multiplication for Negative Numbers." *Arithmetic Teacher 13* (March 1966): 213–217.

Gardner, Martin. "The Concept of Negative Numbers and the Difficulty of Grasping It." *Scientific American* (June 1977): 131–135.

Gibbs, Richard A. "Holes and Plugs." *Mathematics Teacher 70* (December 1977): 19–21.

Hall, Wayne H. "These Three: Minus, Negative, Opposite." *The Arithmetic Teacher 22* (December 1974): 712–713.

Magnuson, Russell C. "Signed Numbers." *The Arithmetic Teacher 14* (November 1966): 573–575.

Sicklick, Francine P. "Patterns in Integers." *Mathematics Teacher 68* (April 1975): 290–292.

Thompson, Patrick W., and Tommy Dreyfuss. "Integers as Transformations." *Journal for Research in Mathematics Education* (March 1988): 115–133.

Williams, David E. "Activities for Algebra." *Arithmetic Teacher 33* (February 1986): 42–47.

Answers

Answers to Instructional Assessment

1. a) Above
   b) 10 hours before
   c) East
   d) Win
   e) Sell
   f) Behind 5 spaces
   g) Spend money
   h) Clockwise
   i) Lose 3 pounds
   j) Retreat/Go back
   k) 1 mile south
   l) Deposit $5
   m) Stretch/Expand

2. a) Minus
   b) Addition
   c) Positive
   d) 3 less than
   e) Positive 8
   f) Right
   g) Decrease
   h) Below

3. b) Lose
   c) Win
   d) Stay put
   e) Above sea level
   f) Before
   g) Middle
   h) Zero

4. a) $^-8$
   b) $^+9$
   c) $^+7$
   d) 0

e) $^+15$
f) $^+13$
g) $^-3$

5. a) $^+4$
   b) $^-3$
   c) $^-4$
   d) $^-4$
   e) $^-4$
   f) $^-5$
   g) 0

6.

7. c) up 5

d) up 9

e) down 15

f) up 8

g) up 8

h) down 8

8. c) $^+5$
   d) $^+9$
   e) $^-15$
   f) $^+8$
   g) $^+8$
   h) $^-8$

9. a) yes
   b) below
   c) 2 degrees below 39
   d) 37 degrees

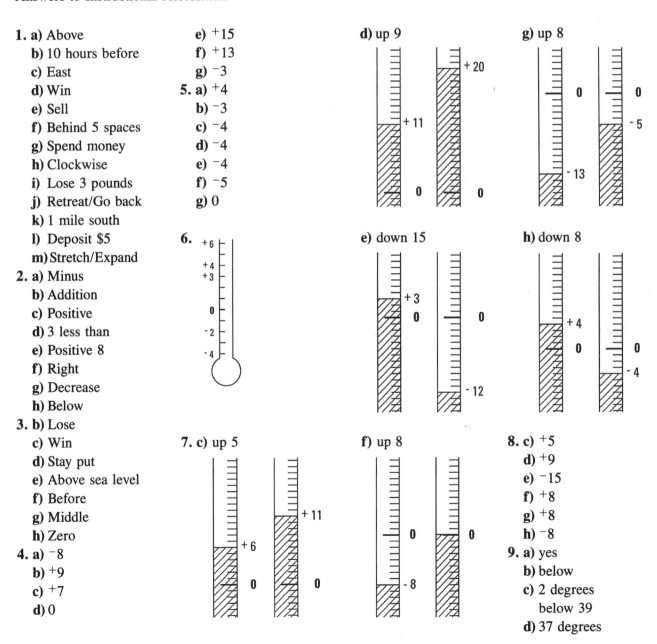

73

10. a) yes
    b) above
    c) 4 degrees
       above ⁻14
    d) ⁻10 degrees

11. a) down
    b) up
    c) below
    d) 4

12. a) ⁺9
    b) ⁺5
    c) ⁻1
    d) ⁺4
    e) ⁻6
    f) ⁺5
    g) ⁻3
    h) ⁻2
    i) ⁻7

13. a) ⁺6
    b) ⁻4
    c) ⁻3
    d) ⁺5
    e) ⁻3
    f) ⁺6
    g) ⁺6
    h) ⁻6

14. a) ⁺3
       ⁻3
       Latreese
    b) ⁻5
       ⁺5
       Jeremy
    c) ⁻1
       ⁺3
       Yolanda

15. a) ⁺2
       ⁺3
       Brian
    b) ⁻4
       ⁻4
       Neither
    c) ⁻3
       ⁻4
       Eloiza
    d) 0
       0
       Neither

16. a) *S*
    b) *H*
    c) *W*
    d) *D*
    e) *E*

17. a)

CPATR number line with arrows pointing to marks:
C P A T R
⁻6 ⁻5 ⁻4 ⁻3 ⁻2   -1   0   +1   ⁺2 ⁺3 ⁺4 ⁺5 ⁺6

    b) *T*
    c) *P*
    d) *A*
    e) *C*

18. a) yes
    b) yes
    c) yes
    d) no
    e) yes
    f) no
    g) yes

Answers to Practice Sheets

Practice sheet 1
1. 2 degrees above zero, ⁺2
2. 4 degrees below zero, ⁻4
3. 5 degrees above zero, ⁺5
4. zero, 0
5. 3 degrees above zero, ⁺3
6. 2 degrees below zero, ⁻2

Practice sheet 2
1. ⁺5

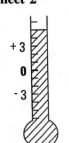

2. ⁺1

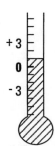

3. ⁻4

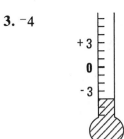

4. ⁻4

5. ⁻3

6. ⁺5

Practice sheet 3
1. 1 unit to the left of zero, ⁻1
2. zero, 0

3. 5 units to the right of zero, ⁺5
4. 2 units to the left of zero, ⁻2
5. 6 units to the left of zero, ⁻6
6. 8 units to the right of zero, ⁺8
Practice sheet 4
1. ⁺7

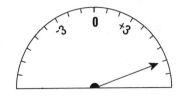

2. ⁻3

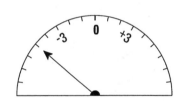

3. ⁻5

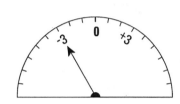

4. ⁺3

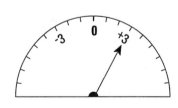

5. ⁺3

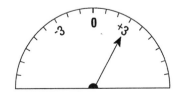

6. ⁻7

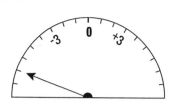

Practice sheet 6
Answers may vary.
For example,
1. 2 degrees above zero, ⁺2

2. 6 degrees below zero, ⁻6

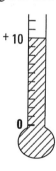

3. 10 degrees above zero, ⁺10

75

4. 0 degrees, 0

5. 10 degrees below zero, ⁻10

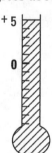

6. 5 degrees above zero, ⁺5

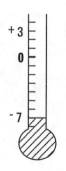

Practice sheet 7
Note: Thermometer scales may vary.
1. ⁻7

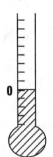

2. 0

3. ⁺3

4. ⁺9

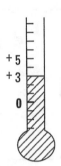

5. ⁺5

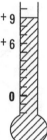

6. ⁻4

Practice sheet 10
Answers may vary.
For example,
1. $\frac{1}{3}$ unit to the left of zero, $-\frac{1}{3}$

2. $\frac{1}{4}$ unit to the right of 1, $+1\frac{1}{4}$

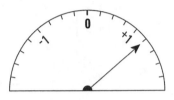

3. $\frac{2}{3}$ unit to the right of ⁻2, $-1\frac{1}{3}$

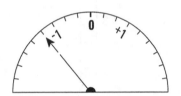

4. 0, 0

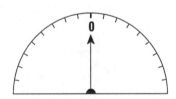

5. $\frac{2}{3}$ unit to the right of zero, $+\frac{2}{3}$

6. $\frac{1}{4}$ unit to the left of 2, $^+1\frac{3}{4}$

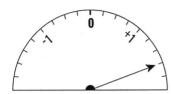

5. $^+5\frac{1}{3}$

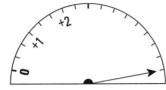

4. 0, 0

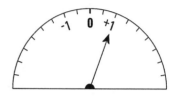

5. 1 unit to the right of zero, $^+1$

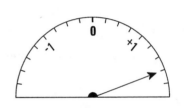

Practice sheet 11
Note: Scales may vary.
1. $^+1\frac{1}{2}$

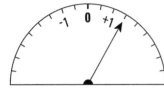

6. $^-2\frac{1}{2}$

Practice sheet 12
Answers may vary.
For example,
1. .25 unit to the left of zero, $^-.25$

6. .75 unit to the right of 1, $^+1.75$

2. $^-2\frac{1}{2}$

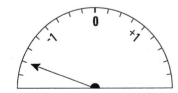

2. .75 unit to the left of 2, $^+1.25$

Practice sheet 13
Note: Scales may vary.
1. $^+1.5$

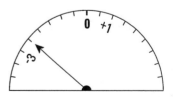

3. $^-3\frac{1}{2}$

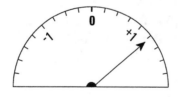

3. .5 unit to the left of zero, $^-.5$

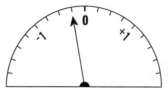

2. $^-2.5$

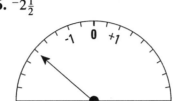

4. $^-1\frac{3}{4}$

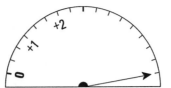

3. ⁻1.75

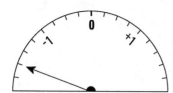

5. ⁺1.5

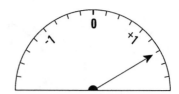

4. ⁻1.5

6. ⁻2.5

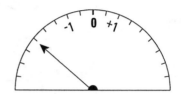

Answers to Follow-up Assessment

1. a) Below

   b) 10 hours after

   c) West

   d) Lose

   e) Buy

   f) Ahead 5 paces

   g) Earn money

   h) Counterclockwise

   i) Gain 3 pounds

   j) Go ahead

   k) 1 mile north

   l) Withdraw $5

   m) Shrink

2. a) Plus

   b) Subtraction

   c) Negative

   d) 3 more than

   e) Negative 8

   f) Left

g) Increase

h) Above

3. a) Tomorrow

   b) Break even

   c) Lose

   d) Go back

   e) Sea level

   f) Now

   g) Left

   h) Positive number

4. a) ⁻8

   b) ⁺2

   c) ⁻7

   d) 0

   e) ⁺6

   f) ⁺7

   g) ⁻1

5. a) ⁺5

   b) ⁺5

c) ⁺4

d) 0

e) ⁻3

f) ⁻2

g) ⁺4

6.

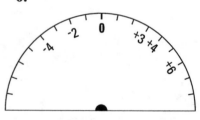

7. a) ⁺5

   b) ⁻6

   c) ⁻2

   d) ⁺8

   e) ⁻4

   f) ⁺7

g) $^+7$

h) $^-7$

8. a) left, 5

    b) right, 6

    c) right, 2

    d) right, 10

    e) left, 2

    f) left, 6

    g) right, 3

    h) left, 8

9. a) $^-5$

    b) $^+6$

    c) $^+2$

    d) $^+10$

    e) $^-2$

    f) $^-6$

    g) $^+3$

    h) $^-8$

10. a) $^+12$

    b) $^+4$

    c) $^-1$

    d) $^-5$

    e) $^+3$

f) $^-5$

g) 0

h) $^-4$

i) $^-7$

11. Order of + and − may vary.

    a) Benita [+] [+] [+] [+] [−] [−]

        Todd [−] [−] [−] [−] [+] [+]

    b) Kim [−] [−] [−] [−] [−] [−]

        Calvin [+] [+] [+] [+] [+] [+]

    c) Renee [+] [+] [+] [−] [−] [−]

        Norm [+] [+] [+] [+] [−] [−]

12. Order of + and − may vary.

    a) Darla [+] [+] [+] [+] [+] [−] [−]

        Tim [+] [+] [+] [+] [+] [−]

    b) Phyllis [−] [−] [−] [−] [+]

        Luis [−] [−] [−] [−] [−] [+] [+]

    c) Maria [−] [−] [+]

        Jack [−] [−] [−] [−] [+] [+]

    d) Florence [+] [+] [−] [−]

        Nate [+] [+] [+] [−] [−] [−]

13. a) E

    b) H

    c) W

d) D

e) E

14. a)

b) A

c) P

d) R

e) R

15. a) no

    b) no

    c) yes

    d) no

    e) no

    f) yes

ALGE**BRIDGE**™

Pattern

Recognition

and

Proportional

Reasoning

Pattern Recognition and Proportional Reasoning

Introduction

Overview. The concept of function is one that is pervasive in mathematics—from simple multiplication tables to complex algebraic equations. The Pattern Recognition and Proportional Reasoning unit is intended to help the student to gain an intuitive sense of the concept of function. If students learn to recognize and generalize numerical patterns, the concept of ratio and the ability to use proportional reasoning can be developed. Once a student understands the meaning of ratio and the idea that many pairs of numbers can be generated from a single fixed ratio, these number pairs (which form numerical patterns) can be used to build the concept of function. By generalizing these numerical patterns, the student can identify rules for functions on an informal level.

The ability to recognize a rule for a given set of number pairs, explain the rule in words, and express the rule symbolically within the domain of nonnegative numbers is developed by having students

- recognize simple geometric and numerical patterns, explain them in words, and generalize the patterns;

- understand and be able to recognize the terminology and notation associated with ratio, including part-to-whole and part-to-hundred relationships;

- use proportional reasoning (equivalent ratios) to solve problems;

- express rules in words;

- express rules in symbols.

Direct variation functions, which logically follow the concept of equal ratios, are treated in this unit. The highest level of comprehension that is addressed in this unit is the generalization of a rule for a given set of number pairs.

Specifics. The ability to **recognize patterns and generalize them** to new problems is critical in mathematics. In this unit, the students are exposed to patterns that help them to develop the type of reasoning that is necessary to formulate a rule for a function in symbolic terms. Although many exercises in pattern recognition may have some inherent instructional value, only those that can be tied to numerical reasoning and equations will be presented here.

The following example illustrates how patterns can lead directly to numerical relationships.

If the pattern of four squares shown in the figure above is continued, how many dots will comprise the 5th square? (Answer: 25)

This type of problem allows the students to see the pattern at a representational level and to use the physical model to generate a hypothesis, so that eventually an extension to a generalized case can be made. To explain how to find the total number of dots in any square in the sequence, the students must see that they multiply the number that is associated with the position of the square in the sequence by itself (*e.g.*, the 8th square will have 8×8 or 64 dots).

In addition, using patterns as a method of instruction gives students the opportunity to talk to one another about mathematics on an informal level. It also helps teachers to realize that they can and should encourage more than one explanation from students of how a particular pattern is generated. It is hoped that these explanations will help students to see that mathematics is a creative subject and not one in which one, and only one, method of solution is possible.

The **concept of ratio** is one that is easier to understand in context than in the abstract. Since this concept readily lends itself to applications, the use of settings that are familiar to students at this level will help them to identify more easily with the situation. As a result, they will be both more interested in the problem and more able to focus on the problem itself.

The development of this concept begins with the representation of ratio on geometric and numerical levels. The student must be able to work with various types of notation that represent ratios as well as to express ratios in different ways.

Once the notation and representation of ratio have been developed, applications may be expanded to include part-to-hundred and part-to-whole relationships. These extensions are crucial in understanding the concept of ratio but are often not given sufficient attention in textbooks. When they are, many students still have trouble grasping the concept because they lack an understanding of the basic notational and representational aspects of ratio. In addition, the part-to-hundred idea is a natural transition to the concept of percent.

Finally, language development plays a major role in building understanding of this concept. Words such as "each," "for every," "per," and "ratio," to name a few, should be used by and discussed with the students. The use and understanding of these words in the context of ratio problems is often overlooked but should be emphasized.

Note that in this unit, a ratio is defined as a type of expression that can be written as a quotient of two numbers, but is denoted without units. For example, the ratio of 6 inches to 1 foot is 1/2, which says that 6 inches is 1/2 as much as 1 foot.

A rate is another type of expression that also can be written as a quotient of two numbers with their respective units, such as kilometers per hour or dollars per item (the denominator is often 1 when two numbers are specified as a rate).

Ratio describes a *size* relationship between two quantities of the same units (*e.g.*, English units of length); rate describes a *unit* relationship between two quantities of different units.

To develop **proportional reasoning**, the idea of ratio is extended to equivalent ratios. Equivalent ratios lead to work with proportions, including how to represent equivalent proportions symbolically and how to determine whether a particular problem contains enough information to use a proportional relationship for its solution. Proportional reasoning is closely related to the understanding of such topics in mathematics as equivalent fractions and percent, as well as many types of functions.

For example, starting with two numbers that are in the ratio 2 to 5 or 2/5, students could generate different number pairs using fractions that are equivalent to 2/5 (*e.g.*, 4/10, 6/15). Another application of proportional reasoning uses the part-to-hundred idea to develop the concept of percent. The number 2/5 can be converted to a percent by re-expressing the denominator as 100 and the numerator as 40. In other words, if 5 is increased to 100, then 2 must be increased to 40 so that the result of the increases remain proportional. Since 2/5 = 40/100 (or 40 per 100), 2/5 of a quantity is the same as 40% of that quantity. Use of the phrase "40 per 100" helps the students with the idea of percent and gives another opportunity to use the word "per." The important point is that proportional reasoning is directly related to equivalent fractions, equivalent ratios, and percent.

Proportional reasoning can also be used to determine the rule for certain types of functions.

Row A	2	3	5	7	12
Row B	18	27	45	63	108

For example, in the table above, each number in row B is 9 times the corresponding number in row A. Although the term "constant of proportionality" need not be used with students at this level, the idea that the ratio of the number in row A to the number in row B remains 1 to 9 throughout is very important in that it relates functions to earlier ideas in this unit. (If $\square$ stands for a number in row A and $\triangle$ stands for the corresponding number in row B, then one can write $9 \cdot \square = \triangle$.)

Proportional reasoning is usually taught in the abstract, but it has applications in many settings that are familiar to the student. It is also prevalent in many topics in algebra, including slopes of linear equations, mixture problems, distance problems, and work problems.

The ability to **express numerical relationships with words** is developed by having students look at patterns or ratios in tables or physical situations and formulate or identify those numerical relationships in words. The development begins with one-step rules and moves into two-step rules. One-step rules are rules that are defined by a single operation and two-step rules are rules that are defined by two operations.

For example, suppose that the student is given the following table and asked to give the rule that relates the numbers in row A to the corresponding numbers in row B.

| **Row A** | 3 | 5 | 6 | 9 | 11 |
| **Row B** | 9 | 15 | 18 | 27 | 33 |

One response could be that each number in row A is 1/3 the corresponding number in row B. Another response could be that each number in row B is 3 times the corresponding number in row A. Activities such as this help the student learn to generalize a numerical relationship in words, so that when these relationships are encountered symbolically in the future study of mathematics, they will be more easily understood. The student's ability to communicate a particular equation or rule for a function in words to another person can greatly enhance the student's understanding of the equation.

Cross-References to
Pre-Algebra and Algebra
Textbook Topics

What follows is a compilation of chapter titles and index entries that are found in typical pre-algebra or algebra textbooks. The concepts and abilities addressed in this unit could be used to enhance the instruction on these topics. (*Note:* Only nonnegative numbers are used in this unit.)

ratio, proportion, and percent	**functions and graphs**
fractions	**rational expressions**
similarity	**variables and expressions**
variables	**linear equations**
number properties and equations	**linear inequalities**
integers and equations	**variation**
formulas (geometry)	

List of Terms

The following is a list of words and phrases that are found in this unit. The teacher may want to add other words and phrases to this list.

add	**is less than**	**representation with figures**
average	**length**	
consecutive (whole numbers)	**less than**	**representation with numbers**
	multiply	
corresponding	**odd number**	**rule**
cube	**pattern**	**sequence**
decrease	**per**	**square**
divide	**percent**	**subtract**
equation	**proportion**	**times as many as**
even number	**ratio**	**times as much as**
for each	**rectangle**	**triple**
for every	**relationship (between numbers)**	**width**
increase		

Classification by Concept

Target Concepts and Abilities	Questions in Instructional Assessment	Questions in Follow-up Assessment	Practice Sheets
Recognize and generalize patterns	1, 2, 13	1, 2, 11	1, 6
Use ratios to express relationships between quantities	3, 4, 5, 7, 8, 14	3, 4, 5, 6, 12	2, 3, 4, 7
Use proportional reasoning	6, 9, 10, 11, 12, 20	7, 8, 9, 10, 18	3, 6, 7, 9
Express rules in words	15, 17, 19	13, 15, 17	5, 6, 8, 9
Express rules in symbols	16, 18, 21	14, 16, 19	5, 6, 8, 9

Classification by Question

Target Concepts and Abilities		Recognize and generalize patterns	Use ratios to express relationships	Use proportional reasoning	Express rules in words	Express rules in symbols
Question Number or Practice Sheet Number	1	IA FA PS				
	2	IA FA	PS			
	3		IA FA PS	PS		
	4		IA FA PS			
	5		IA FA		PS	PS
	6	PS	FA	IA PS	PS	PS
	7		IA PS	FA PS		
	8		IA	FA	PS	PS
	9			IA FA PS	PS	PS
	10			IA FA		
	11	FA		IA		
	12		FA	IA		
	13	IA			FA	
	14		IA			FA
	15				IA FA	
	16					IA FA
	17				IA FA	
	18			FA		IA
	19				IA	FA
	20			IA		
	21					IA

IA - Questions in Instructional Assessment
FA - Questions in Follow-up Assessment
PS - Practice Sheets

88

Instructional Assessment

The questions in both the Instructional and Follow-up Assessment Instruments for this unit have different formats appropriate to the nature of the concepts addressed in the questions.

It is necessary to introduce students to these formats since they may have limited experience with different types of responses.

If there is not sufficient time to review examples of the different question types, review the directions below with the students before they work the *Algebridge*™ exercises.

Directions

There are three types of questions in this exercise.

One type asks you to choose the answer to the problem and mark the corresponding letter, A, B, C, or D, on your answer sheet.

The second type gives you several choices related to a specific question. You must answer **yes** or **no** for each of these choices.

The third type asks you to write your own answer to the problem.

Focus of Question 1

In question 1 students use a geometric representation of a number pattern to determine the next figure in the pattern. Although the primary focus is on the *number* of triangles that should be in the next figure, the orientation of the last triangle is also considered. Thus, this question addresses both numerical and geometric thinking.

Question 1

1. The figures below form a pattern. Draw the next figure in the pattern.

  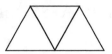 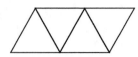 ?
___

Answer: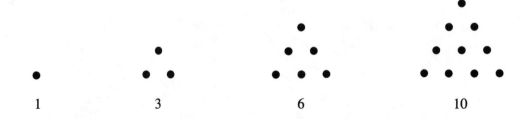

Guiding Class Discussion

Students should understand that there should be five triangles in the next picture. In discussing this problem with the students, the teacher could ask, "What would the 25th figure look like?" One way to solve this problem is to realize that

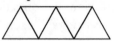

is the next figure; therefore, the 25th figure would be comprised of 25 triangles arranged as follows:

It is important to note the number of triangles in the 25th figure as well as the orientation of the last triangle. In discussing the orientation of the last triangle, students could be encouraged to make a generalization about the representations for even versus odd numbers.

The practice sheets and suggested activities in the Follow-up Instruction section under the concept **Recognize and generalize patterns** may be helpful for those students having difficulty or needing more practice.

Focus of Question 2

This question allows the teacher to assess the student's ability to discriminate between geometric (or representational) and numerical thinking.

Question 2

2.

<table>
<tr><td>•</td><td>•</td><td>•</td><td>•</td></tr>
<tr><td>1</td><td>3</td><td>6</td><td>10</td></tr>
</table>

Shown in the figure are the first four triangular numbers. What is the next triangular number?

(A) 12
(B) 15
(C) 18
(D) 20

Guiding Class Discussion

Have the student discuss *how* he or she solved the problem. Ask questions to determine whether the student got the answer geometrically or numerically. Emphasizing that both approaches are valid reinforces the idea that a problem may often be solved in more than one way. The discussion may also provide the teacher with clues about instructional approaches that may be more meaningful for particular students.

It may be helpful to point out to the student that the difference between successive numbers in the sequence increases by one ($3 - 1 = 2$, $6 - 3 = 3$, $10 - 6 = 4$) and that this difference could be used to find the next number.

The practice sheets and suggested activities in the Follow-up Instruction section under the concept **Recognize and generalize patterns** may be helpful for those students having difficulty or needing more practice. In particular, **practice sheet 1** presents more questions that focus on observing and using geometrical and numerical patterns.

Focus of Question 3

Question 3 introduces the terminology and symbolism that are related to the concept of ratio.

Question 3

Questions 3–5 refer to the following situation.

Manny sold 2 lemon ices for every 5 cherry ices.

3. Consider each of the following. Is the ratio of the number of lemon ices to the number of cherry ices correctly expressed?

(yes)	no	**a)**	2 to 5
yes	(no)	**b)**	2 equals 5
(yes)	no	**c)**	$\dfrac{2}{5}$
yes	(no)	**d)**	2 less than 5
(yes)	no	**e)**	$2 : 5$
(yes)	no	**f)**	Two for each five
yes	(no)	**g)**	Two times as much as five

Analysis of Question 3

The student should be able to identify all valid representations for the ratio in this context and to realize that a ratio can be expressed in many ways.

It may be that no one will answer **yes** to 3f. Students who do should be questioned to see whether they understood the concept or just guessed. This question represents the most sophisticated interpretation of the concept of ratio for students at this level.

Guiding Class Discussion

The practice sheets and suggested activities in the Follow-up Instruction section under the concept **Use ratios to express relationships between quantities** may be helpful for those students having difficulty or needing more practice.

Focus of Question 4

Question 4 introduces the idea of equivalent ratios.

Question 4

Questions 3–5 refer to the following situation.

Manny sold 2 lemon ices for every 5 cherry ices.

4. Consider each of the following. Could the numbers represent the number of lemon ices and cherry ices that Manny sold?

			Lemon	Cherry
(yes)	no	a)	2	5
yes	(no)	b)	5	2
yes	(no)	c)	5	7
(yes)	no	d)	10	25
yes	(no)	e)	10	2
yes	(no)	f)	25	10
yes	(no)	g)	10	4

Analysis of Question 4

In discussing the problem, have students set up the ratio 2 to 5 as the fraction $\frac{2}{5}$, then investigate whether $\frac{2}{5}$ is equal to the fractional form of each of the choices; this exercise will reinforce the concept of equivalent fractions. Choices 4f and 4g may present an opportunity to discuss the *direction* of the ratio, which is also dealt with in question 8.

Guiding Class Discussion

The practice sheets and suggested activities in the Follow-up Instruction section under the concept **Use ratios to express relationships between quantities** may be helpful for those students having difficulty or needing more practice.

Focus of Question 5

Question 5 assesses whether students can use the figure to break up the number 14 into two parts that are in a 2 to 5 ratio.

Question 5

Questions 3–5 refer to the following situation.

Manny sold 2 lemon ices for every 5 cherry ices.

5. Each 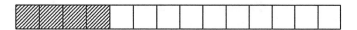 stands for one lemon ice that Manny sold and each ☐ stands for one cherry ice that he sold.

    a)   If Manny sold only lemon and cherry ices, shade the correct number of ☐'s in the figure below to show how many out of 14 ices sold were lemon.

    b)   How many ☐'s did you shade? Answer: _4_

Analysis of Question 5

In algebra, the ability to decompose a given quantity into two or more quantities according to a specified ratio is important. The context of this question should help the students see that often a figure may be useful in solving a problem such as this.

Practice sheet 2 focuses on this concept and provides the student with an opportunity for additional practice.

Guiding Class Discussion

The practice sheets and suggested activities in the Follow-up Instruction section under the concept **Use ratios to express relationships between quantities** may be helpful for those students having difficulty or needing more practice.

Focus of Questions 6–7

In questions 6 and 7, students use the relationship between the radius and diameter of a circle to complete a table and write a ratio.

Questions 6–7

Questions 6–7 refer to the following figures.

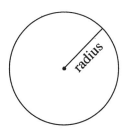

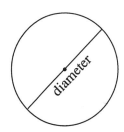

6. The table below shows different lengths for the radius and diameter of a circle. Fill in all missing numbers in the table.

Row A	Length of Radius	1	2	3	4	5	_8_	_15_	_24_
Row B	Length of Diameter	_2_	_4_	_6_	_8_	_10_	16	30	48

7. For each pair of numbers in the table in question 6, what is the ratio, in simplest form, of the length of the radius to the length of the corresponding diameter?

Answer: _1 to 2_

Analysis of Questions 6–7

For those students who may be unfamiliar with the terms "radius" and "diameter," illustrations are presented in question 6. After the table in question 6 has been completed, question 7 permits students to demonstrate which notation for representing a ratio is most natural to them.

In the discussion of question 6, the teacher should point out to the students that the relationship that exists between each number in row A and the corresponding number in row B can be described by the ratio in question 7.

Guiding Class Discussion

The practice sheets and suggested activities in the Follow-up Instruction section under the concepts **Use ratios to express relationships between quantities** and **Use proportional reasoning** may be helpful for those students having difficulty or needing more practice.

Focus of Questions 8–12

Like question 3, question 8 assesses the student's knowledge of whether a given notation is correct for representing a ratio. In addition, it assesses whether the student recognizes the distinction between the ratio of 5 to 7 and the ratio of 7 to 5. Question 9 assesses the idea of proportional reasoning through equivalent ratios. The purpose of question 10 is to ascertain whether the students realize what information is needed to determine one or more of the values in a problem that involves proportional reasoning. Questions 11 and 12 introduce the use of a proportion.

Questions 8–12

Questions 8–12 refer to the following situation.

For a school play, a ticket office sold 5 reserved tickets for every 7 general admission tickets.

8. Consider each of the following. Is the ratio of the number of reserved tickets to the number of general admission tickets correctly expressed?

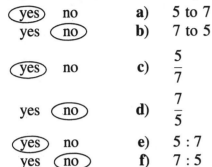

  ⓨes no **a)** 5 to 7

  yes ⓝo **b)** 7 to 5

  ⓨes no **c)** $\dfrac{5}{7}$

  yes ⓝo **d)** $\dfrac{7}{5}$

  ⓨes no **e)** 5 : 7

  yes ⓝo **f)** 7 : 5

9. Consider each of the following. Is the ratio of the number of reserved tickets to the number of general admission tickets the same as 5 to 7?

  ⓨes no **a)** 10 reserved for every 14 general admission

  yes ⓝo **b)** 6 reserved for every 8 general admission

  yes ⓝo **c)** 28 reserved for every 20 general admission

  yes ⓝo **d)** 11 reserved for every 15 general admission

  ⓨes no **e)** 15 reserved for every 21 general admission

  yes ⓝo **f)** 500 reserved for every 70 general admission

10. Consider each of the following facts. If 5 reserved tickets were sold for every 7 general admission tickets, would knowing the fact help you to determine exactly how many general admission tickets were sold?

  ⓨes no **a)** The number of reserved tickets sold

  ⓨes no **b)** The total number of reserved and general admission tickets sold

  yes ⓝo **c)** The cost of 1 general admission ticket

  yes ⓝo **d)** The total amount collected from ticket sales

11. If 100 reserved tickets were sold, how many general admission tickets were sold?

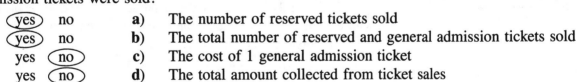

  Answer: _140_

12. Let □ stand for the number of general admission tickets sold. Consider each of the following. If 100 reserved tickets were sold, could the proportion be used to find the number that goes in the □ ?

yes ~~(no)~~   **a)** $\dfrac{100}{7} = \dfrac{5}{\square}$

~~(yes)~~ no   **b)** $\dfrac{100}{\square} = \dfrac{5}{7}$

yes ~~(no)~~   **c)** $\dfrac{100}{5} = \dfrac{7}{\square}$

yes ~~(no)~~   **d)** $\dfrac{100}{7} = \dfrac{\square}{5}$

~~(yes)~~ no   **e)** $\dfrac{100}{5} = \dfrac{\square}{7}$

Analysis of Questions 8–12

In question 8, all symbols used are correct for representing ratios.

In question 9 students should be encouraged to express each choice in fractional form.

Question 10 could be used in a class discussion of abundant versus deficient data, focusing first on problems that can be solved using a proportion and leading to a general discussion of sufficient data needed to solve problems.

Question 11 is given in a free-response format. Students may be able to "figure out" the solution to the problem but may not know what to do with the abstract representation in question 12. Class discussion of this question should probe to find out how the students solved the problem.

Question 12 introduces the notation used to state a proportion with a placeholder in the position of the unknown value. A discussion of this problem should include the fact that a proportion can be written in different ways and still yield the same result.

Guiding Class Discussion

The practice sheets and suggested activities in the Follow-up Instruction section under the concepts **Use ratios to express relationships between quantities** and **Use proportional reasoning** may be helpful for those students having difficulty or needing more practice.

In particular, **practice sheet 3** can be used to discuss equivalent ratios, the use of proportions for solving problems, and part-to-whole relationships. **Practice sheet 4** is particularly helpful for emphasizing part-to-whole relationships and percent.

Focus of Questions 13–16

Questions 13–16 involve the relationship between the length of the side of a square and the perimeter. Knowing the definition of "perimeter" is not required to answer any of the questions.

Questions 13–16

Questions 13–16 refer to the following figures.

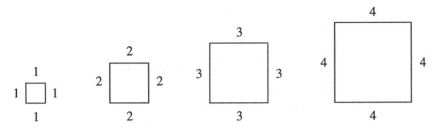

Each square is formed by segments whose lengths are shown.

13. Use the squares above to fill in the table below.

	Column A (Length of a Side)	**Column B** (Total Length of the 4 Segments in That Square)
a)	*1*	*4*
b)	*2*	*8*
c)	*3*	*12*
d)	*4*	*16*

14. For each square above, what is the ratio, in simplest form, of the length of a side to the total length of the 4 segments in that square? (The total length of the 4 segments is called the perimeter of the square.)

Answer: *1 to 4*

15. Consider each of the following. Can the rule be applied to each number in column A in the table in question 13 to get the corresponding number in column B?

yes	(no)	**a)**	Add 3 to the number in column A.
yes	(no)	**b)**	Add 4 to the number in column A.
(yes)	no	**c)**	Multiply the number in column A by 4.
yes	(no)	**d)**	Divide the number in column A by 4.

16. Let □ and △ stand for two numbers such that $4 \cdot \square = \triangle$. Consider each of the following. Is the statement true?

(yes)	no	**a)**	If 0 is placed in the □, then 0 should be placed in the △.
(yes)	no	**b)**	If 1 is placed in the □, then 4 should be placed in the △.
yes	(no)	**c)**	If 2 is placed in the □, then 6 should be placed in the △.
yes	(no)	**d)**	If 4 is placed in the □, then 1 should be placed in the △.
(yes)	no	**e)**	If 4 is placed in the □, then 16 should be placed in the △.

Analysis of Questions 13–16   After the table in question 13 has been completed, students can use those numbers to determine the ratio in question 14.

Question 15 asks the students to identify in words the rule that describes the relationship between columns A and B in the table. The teacher should emphasize to the students that whenever a rule is generalized, it must apply to all numbers involved. In this case, 15a applies to the first pair of numbers only, so it cannot be the rule, but 15c applies to all number pairs in the table, so it must be a correct representation of the rule.

Question 16 shows students that equations in two variables can be satisfied by more than one number pair. Thus students should realize that multiple number pairs can be generated from a rule and that one number in each pair is dependent on the other. Also this question provides practice for replacement techniques with whole numbers. To extend the concept, fractions and decimals could be used. The aid of a calculator may be helpful.

Guiding Class Discussion

The practice sheets and suggested activities in the Follow-up Instruction section under the concepts **Recognize and generalize patterns**, **Use ratios to express relationships between quantities**, **Express rules in words**, and **Express rules in symbols** may be helpful for those students having difficulty or needing more practice.

Focus of Questions 17–18

Questions 17–18 extend development of the ability to express numerical relationships with words to expression with symbols.

Question 17 involves identifying a one-step rule that describes what is occurring in the table. The student should be able to identify multiple valid descriptions of the rule.

Question 18 assesses whether a student can generalize a rule and express the generalization in symbols.

Questions 17–18

Questions 17–18 refer to the following figures.

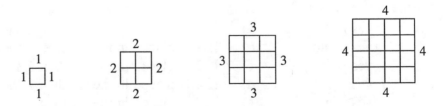

The area of each square above is shown in the table below.

Column A (Length of a Side)	Column B (Area)
1	1
2	4
3	9
4	16

17. Consider each of the following rules. Can the rule be applied to each number in column A in the table to get the corresponding number in column B?

yes (no) **a)** Multiply the number in column A by 2.
(yes) no **b)** Multiply the number in column A by itself.
yes (no) **c)** Add 2 to the number in column A.
(yes) no **d)** Square the number in column A.

18. Let □ stand for a number in column A and △ stand for the corresponding number in column B. Consider each of the following. Could it be the rule in symbols?

yes (no) **a)** $2 \cdot \square = \triangle$
(yes) no **b)** $\square \cdot \square = \triangle$
yes (no) **c)** $2 + \square = \triangle$
yes (no) **d)** $\square + \square = \triangle$

Analysis of Questions 17–18

In question 17 both choices b and d are correct. This type of problem should illustrate to the student that there may be more than one correct statement of a given rule.

Students should acquire the habit of describing the rule in words before they attempt to use symbols in these types of problems. Students may wish to apply the technique learned in question 16 (substituting values for □) to help identify the rule in symbols.

Question 18 also provides a good opportunity for discussing 2 as a special number in that $2 + 2 = 2 \times 2$. The only other number for which this is true is 0.

Guiding Class Discussion

The practice sheets and suggested activities in the Follow-up Instruction section under the concepts **Express rules in words** and **Express rules in symbols** may be helpful for those students having difficulty or needing more practice.

Focus of Questions 19–21

Question 19 is more difficult than question 15 since it involves formulating a *two*-step rule in words.

In question 20, students are asked to apply the rule by completing the table, and in question 21, they must identify the rule in symbols.

Questions 19–21

Questions 19–21 refer to the following table.

Gina's Running Record

Column A (Number of Days)	Column B (Total Distance Run in Miles)
2	8
3	10
4	12
5	14

19. Consider each of the following rules. Can the rule be applied to each number in column A in the table to get the corresponding number in column B?

    yes (no)   **a)**   Take 5 times the number in column A and subtract 2.

    (yes) no   **b)**   Take 2 times the number in column A and add 4.

    yes (no)   **c)**   Take 6 times the number in column A and subtract 4.

    yes (no)   **d)**   Cube the number in column A.

    (yes) no   **e)**   Add 2 to the number in column A and then double the result.

20. If the trend in the table continued through day 10, fill in the missing numbers in the table.

Gina's Running Record

Column A (Number of Days)	Column B (Total Distance Run in Miles)
2	8
3	10
4	12
5	14
6	*16*
10	*24*

21. Let □ stand for a number in column A and △ stand for the corresponding number in column B. Consider each of the following. Could it be the rule in symbols?

    yes (no)   **a)** $(5 \cdot \square) - 2 = \triangle$

    (yes) no   **b)** $(2 \cdot \square) + 4 = \triangle$

    yes (no)   **c)** $(6 \cdot \square) - 4 = \triangle$

    yes (no)   **d)** $\square \cdot \square \cdot \square = \triangle$

    (yes) no   **e)** $2 \cdot (\square + 2) = \triangle$

Analysis of Questions 19–21   The fact that 19b and 19e are both valid descriptions for the rule illustrates that there is more than one way to formulate a function's rule in words. However, it is more important that students understand that each description must be applicable to every corresponding pair of numbers in the given problem.

Choice e in question 21 illustrates that $2 \cdot \square + 4$ is equivalent to $2 \cdot (\square + 2)$. This should be discussed with the students. The teacher should point out to students who answered **yes** to 21a, 21c, or 21d that it is necessary to check all number pairs, not just the first number pair.

Guiding Class Discussion

The practice sheets and suggested activities in the Follow-up Instruction section under the concepts **Use proportional reasoning, Express rules in words**, and **Express rules in symbols** may be helpful for those students having difficulty or needing more practice.

In particular, **practice sheets 5–9** provide more opportunities for discussion and practice in expressing numerical relationships in words and symbols. Note that **practice sheets 6 and 7** can serve a summarizing function since they also include questions with geometric patterns.

Follow-up Instruction

After the Instructional Assessment has been used to pinpoint students' conceptual weaknesses, the teacher may wish to use the following suggested activities and practice sheets to help correct those misconceptions.

Recognize and generalize patterns

Practice sheet 1 contains questions that involve observing patterns. The first two questions provide the student with an opportunity to reason on a numerical level beyond what is given in the figure. These questions allow the teacher to assess for understanding of multiple operations and for discrimination among operations.

Discussion of these questions could include a discussion of odd and even numbers since the answers depend on whether the row in question is odd- or even-numbered. (The row sum for each odd-numbered row is 2 and for each even-numbered row is 0.) Also, the odd-numbered rows have an odd number of terms and the even-numbered rows have an even number of terms.

The third question encourages reasoning techniques different from those in 1 and 2. It provides a good basis for a discussion of the reasoning process necessary before a pattern can be generalized.

Two different methods for finding the number that replaces □ are (1) observing the pattern in the numbers in the 4-8-16-□ diagonal and (2) multiplying by each other the two numbers on either side of □ in the row above □, namely 2 and 16.

The fourth question presents another opportunity to discuss the fact that there may be different valid ways to complete a sequence. In this question, the next figure will contain 4 hexagons. However, addition of the fourth hexagon could produce several different results. For example,

or

103

Questions 5a and 5b can be used to reinforce understanding of the terms "even," "odd," and "consecutive," but are presented in the context of pattern development. A question that parallels 5b could be developed for even numbers.

The last question on this sheet illustrates that, in some cases, more than one description of a pattern may be valid. For example, one description of this pattern is that starting with 1, add 2, subtract 1, add 2, add 1, and repeat in this sequence. Another description is that the first two consecutive positive odd numbers are followed by the first two consecutive positive even numbers, which are followed by the next two consecutive odd numbers, and so forth. If the student can justify a particular pattern that is applicable in all cases, it should be accepted.

Activity. An activity that would be appropriate at the end of this unit is having students generate their own number patterns.

Practice sheet 6 also addresses this concept as do questions 1, 2, and 11 in the Follow-up Assessment.

Use ratios to express relationships between quantities

Practice sheet 2, similar to question 5 in the Instructional Assessment, requires students to divide figures representing groups of objects into smaller groups according to a given ratio. In answering question 4, students should be asked to explain why the group can or can not be divided according to the given ratio.

Practice sheets 3 and 4 contain questions that address the concepts of ratio and proportional reasoning.

The first question on practice sheet 3 involves the concept of equivalent ratios. Students should be asked whether 10 and $2\frac{1}{2}$ would be a correct answer to the question. This discussion should lead to the realization that even though $10 : 2\frac{1}{2} = 4 : 1$, it is not reasonable to talk about $2\frac{1}{2}$ dogs. Students could then be reminded to check the reasonableness of their answers. Another example such as "How many dogs would Harry have if he had 5 cats?" will emphasize this idea.

The second question on practice sheet 3 assesses a student's ability to identify a correct representation for a specific ratio. The phrase "for each," which is pertinent vocabulary for this concept, is also presented.

The third question assesses understanding of the part-to-whole relationship, which the teacher may extend to work with decimals and percent. (Practice sheet 4 uses figures to provide further practice with part-to-whole and percent.)

The part-to-whole concept is developed on a geometric level using the pieces of the tangram puzzle in the fourth question. The pieces are not connected so that the student will not count the large triangle formed by the two largest individual triangles.

The last question addresses the same ideas as do questions 11 and 12 of the assessment. Students choosing the correct answer could be asked whether they arrived at the answer by setting up a proportion or by using another method. Students choosing C or D should be asked about the reasonableness of their answer.

Practice sheet 7 as well as Follow-up Assessment questions 3–6 and 12 also address this concept.

Use proportional reasoning

While **practice sheets 7 and 9** address several concepts presented in this unit, their primary focus is on using proportional reasoning. For practice sheet 7, students cut rectangles from graph paper to decide if the rectangles fit a given pattern. Students are asked to explain how to determine if any new rectangle is part of the pattern. Scissors and graph paper are needed for this activity, which may be worked on individually or in small groups. Practice sheet 9 presents tables of numbers and asks students to determine and/or use rules that describe the given relationships. Further, students explore the effect on the corresponding table entry when those in Column A are increased by addition or multiplication by a given number.

Note: **Practice sheets 7 and 9** extend the concepts assessed in this unit. These practice sheets are more challenging than those presented previously. The teacher may want to assign one or more of them to the more capable students.

Practice sheet 3 (described under the previous concept Use ratios to express relationships between quantities) and **practice sheet 6** also deal with proportional reasoning.

Follow-up Assessment questions 7–10 and 18 also address this concept.

Express rules in words

Practice sheet 5 contains questions that address the concepts of expressing numerical relationships with words and symbols. These questions lead to the formulation of a rule for a particular function in symbols. The important thing for the student to realize at this point is that a better understanding of symbolic numerical relationships is possible if those relationships can be expressed in words.

The second question can be used to provide practice in formulating and using a two-step rule and in using a calculator to perform multiple arithmetic operations. The teacher should specify how many additional number pairs should be generated.

The questions on this practice sheet lead directly to an understanding of linear equations in two variables.

Practice sheets 6 and 8 are especially useful for practice in expressing rules in words and symbols. Practice sheet 6 presents geometric patterns for description while practice sheet 8 presents familiar real world situations in a verbal context.

Practice sheet 9 also relates to this concept.

Activity. For additional activities, the teacher can consult pre-algebra and algebra textbooks. Sections of the textbooks on graphing that give values in a two-column or two-row format provide the teacher with the basis for a lesson on expressing numerical relationships in words and symbols.

Questions 13, 15, and 17 in the Follow-up Assessment also address this concept.

Express rules in symbols

Practice sheet 5 (discussed under the previous concept Express rules in words) and **practice sheets 6, 8, and 9** involve expressing rules in symbolic form.

Questions 14, 16, and 19 in the Follow-up Assessment also address this concept.

Note: **Practice sheets 6 and 7** can be used at the conclusion of the unit, since they focus on concepts that are stressed throughout.

Bibliography of Resource Materials

Dolan, D., and J. Williamson. *Teaching Problem-Solving Strategies.* Reading, MA: Addison-Wesley, 1983.

Jacobs, Harold. *Mathematics—A Human Endeavor.* New York: W. H. Freeman, 1982.

Lappan, G., et al. *Similarity and Equivalent Fractions.* Reading, MA: Addison-Wesley, 1986.

Post, Thomas R., et al. "Proportionality and the Development of Prealgebra Understandings." *1988 Yearbook of the National Council of Teachers of Mathematics.* Reston, VA: National Council of Teachers of Mathematics, 1988:78–90.

Reeves, Charles. *Problem-Solving Techniques Helpful in Mathematics and Science.* Reston, VA: National Council of Teachers of Mathematics, 1987.

Shroyer, J., and W. Fitzgerald. *Mouse and Elephant: Measuring Growth.* Reading, MA: Addison-Wesley, 1986.

Answers

Answers to Instructional Assessment

1.

2. B

3. a) yes
   b) no
   c) yes
   d) no
   e) yes
   f) yes
   g) no

4. a) yes
   b) no
   c) no
   d) yes
   e) no
   f) no
   g) no

5. a)
   b) 4

6. Row A 8, 15, 24
   Row B 2, 4, 6, 8, 10

7. 1 to 2

8. a) yes
   b) no
   c) yes
   d) no
   e) yes
   f) no

9. a) yes
   b) no
   c) no
   d) no
   e) yes
   f) no

10. a) yes
   b) yes
   c) no
   d) no

11. 140

12. a) no
   b) yes
   c) no
   d) no
   e) yes

13. a) 1, 4
   b) 2, 8
   c) 3, 12
   d) 4, 16

14. 1 to 4

15. a) no
   b) no
   c) yes
   d) no

16. a) yes
   b) yes
   c) no
   d) no
   e) yes

17. a) no
   b) yes
   c) no
   d) yes

18. a) no
   b) yes
   c) no
   d) no

19. a) no
   b) yes
   c) no
   d) no
   e) yes

20. 16
   24

21. a) no
   b) yes
   c) no
   d) no
   e) yes

Answers to Practice Sheets

Practice sheet 1
1. A
2. B
3. B
4. Answers will vary.
   For example,

5. a) B
   b) C
6. B

Practice sheet 2
1.

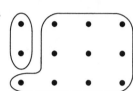

2.

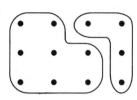

3.

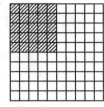

4. no, no

Practice sheet 3
1. a) yes
   b) no
   c) no
   d) no
   e) yes
2. A
3. D
4. A
5. B

Practice sheet 4

1.

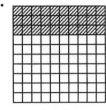

$$\frac{25}{100}$$

2.

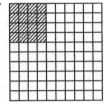

$$\frac{16}{100}$$

3.

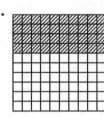

$$\frac{40}{100}$$

4.

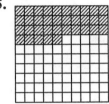

$$\frac{30}{100}$$

5.

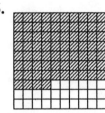

$$\frac{35}{100}$$

6.

$$\frac{74}{100}$$

Practice sheet 5

1. a) Multiply each number in A by 2, then add 3.

   b) $2 \times \square + 3 = \triangle$

   c) 11

   d)

Column A	Column B
9	21
13	29

2. a) Multiply each number in A by 3, then add 2.

   b) Answers will vary. For example, if 37 is in Column A, 113 should be in Column B.

Practice sheet 6

1.
```
•   •
•   •
•   •
•   •
```
Shape 4

2. Multiply shape number by 2.

3.

Number of Shape	Number of Dots in Shape
1	2
2	4
3	6
4	8
5	10
6	12

4. 20, 86

5. $2 \times \square = \triangle$

6. Multiply shape number by 3 and add 2.

7. Answers will vary but pattern should correspond to rule. For example, if the rule is multiply by 2 and add 3, Shape 1 should have 5 dots; Shape 2, 7 dots; Shape 3, 9 dots; Shape 4, 11 dots.

Practice sheet 7

2. 8 units by 16 units; 4

3. Answers will vary. For example, a 2-unit by 3-unit rectangle is not part of the pattern, but a 16-unit

by 32-unit rectangle is part of the pattern.

4. For every rectangle in the pattern, the longer side is twice as long as the shorter side.

5.

Width	Length
$\frac{1}{4}$	$\frac{1}{2}$
$\frac{1}{2}$	1
1	2
2	4
4	8
8	16
16	32
32	64

6. Answers will vary (but should include the idea of length to width ratio of 2 to 1).

Practice sheet 8

1. $2.00 plus ($.50 times the number of hours) equals cost

2. $.50 $\times \square$ + $2.00 = $\triangle$

3. 6

4. ($.20 times the number of tickets) plus $50.00 equals cost

5. $.20 $\times \square$ + $50.00 = $\triangle$

6. 625

Practice sheet 9

1. Increased by 9 times original area.

2. a) Multiply each number in A by 3.

   b)

Column A	Column B
2	6
35	105
8	24

   c) Answers will vary. For example, if 18 is placed in column A, the corresponding number for column B is 54.

   d) Increases by 6.

   e) Increases by 3×28 or 84.

3. 4

Answers to Follow-up Assesment

1. ↑

For example,

Ⓗ Ⓗ Ⓗ Ⓗ Ⓗ Ⓗ
Ⓗ Ⓗ Ⓗ Ⓗ Ⓗ Ⓗ
Ⓗ Ⓗ Ⓗ Ⓗ Ⓗ Ⓗ
Ⓒ Ⓒ Ⓒ Ⓒ Ⓒ
Ⓒ Ⓒ Ⓒ Ⓒ Ⓒ

b) 12

2. B

3. a) no
b) yes
c) no
d) yes
e) no
f) yes
g) no
h) yes

6. a) yes
b) no
c) yes
d) no
e) yes
f) no

4. a) no
b) no
c) yes
d) no
e) no
f) yes
g) no

7. a) yes
b) yes
c) no
d) no
e) no
f) yes

8. a) no
b) yes
c) yes
d) no

5. a) Placement of H's
and C's may vary,
but there should be
21 H's and 12 C's.

9. 28
10. a) no
b) no
c) yes
d) yes
e) no

11.

	Column A	Column B
a)	1	3
b)	2	6
c)	3	9
d)	4	12

12. 3 to 1

13. a) yes
b) yes
c) no
d) no

14. a) yes
b) yes
c) no
d) no

15. a) no
b) yes
c) yes
d) no

16. a) no
b) no
c) no
d) yes

17. a) no
b) no
c) yes
d) no
e) yes

18.

Column A	Column B
9	19
12	25

19. a) no
b) yes
c) no
d) no
e) yes

ALGE**B**|**R**|**I**|**D**|**G**|**E**™

Constructing

Numerical

Equations

Constructing Numerical Equations

Introduction

The focus in this unit is on understanding the relationships among quantities in a problem and on being able to write a numerical equation that represents these relationships. The emphasis on language and symbolization helps students to develop the skills needed to translate the situation given in a word problem into a mathematical statement. The concept of variable is introduced by asking students to consider the effect of using different numbers in the basic relationship between quantities expressed by the problem.

Students who can understand the relationship between quantities presented in a word problem and who can construct a numerical equation that represents these relationships will be well prepared to solve word problems that involve algebra.

In the context of word problems that can be modeled with simple linear equations, this unit develops the students' ability to

- understand the vocabulary used in word problems;

- use mental representations to help understand the situation in the problem;

- pay attention to the details of language and symbolization;

- translate vocabulary into mathematical statements and vice versa;

- understand the concept of variable by introducing placeholders to represent quantities;

- recognize information that is necessary and sufficient to solve the problem.

Cross-References to Pre-Algebra and Algebra Textbook Topics

The concepts and abilities addressed in this unit are not often treated explicitly in pre-algebra and algebra textbooks. Word problems appear throughout the textbooks, however, and the use of this unit would be helpful when students are introduced to word problems.

This unit may also be helpful when studying the following topics that appear in many pre-algebra and algebra textbooks.

algebraic expressions	**variables**
open sentences	**word problems**
problem solving	

List of Terms

In addition to the terminology used in the Instructional Assessment, the following is a list of words and phrases that are frequently used in word problems. The teacher may want to add other words and phrases to this list.

at most, at least	**how many times more**
as many as, times as many as	**how many times less than**
as much as, times as much as	**how many times more than**
consecutive	**less than, is less than**
difference	**more than, is more than**
divisible by	**odd**
even	**older than, younger than**
greater than, less than	**product**
greatest, least	**quotient**
half as many as	**sum**
how many more	**times**

Classification by Concept

Target Concepts and Abilities	Questions in Instructional Assessment	Questions in Follow-up Assessment	Practice Sheets
Understand the vocabulary used in word problems	1, 2, 3, 4, 8, 9, 10, 18	1, 2, 3, 10	1–10
Use mental representations	2	9	1–5
Pay attention to the details of language and symbolization	5, 6, 11, 12, 18, 19, 20, 21, 22	4, 5, 10, 11, 12	1–10
Translate vocabulary into mathematical statements and vice versa	6, 7, 12, 13, 14, 18, 19, 20, 21, 22, 23	5, 6, 7, 10, 11, 12, 13, 14, 15, 16, 17, 18, 20	1–10
Understand the concept of variable by introducing placeholders	7, 13, 14, 21, 22	6, 7, 12, 13, 14, 16, 17, 19	5
Recognize necessary and sufficient information	15, 16, 17, 23	8, 20	11, 12

Classification by Question

Target Concepts and Abilities	Understand the vocabulary used in word problems	Use mental representations	Pay attention to the details of language and symbolization	Translate vocabulary into mathematical statements and vice versa	Understand the concept of variable by introducing placeholders	Recognize necessary and sufficient information
Question Number or Practice Sheet Number 1	IA FA PS	PS	PS	PS		
2	IA FA PS	IA PS	PS	PS		
3	IA FA PS	PS	PS	PS		
4	IA PS	PS	FA PS	PS		
5	PS	PS	IA FA PS	FA PS	PS	
6	PS		IA PS	IA FA PS	FA	
7	PS		PS	IA FA PS	IA FA	
8	IA PS		PS	PS		FA
9	IA PS	FA	PS	PS		
10	IA FA PS		FA PS	FA PS		
11			IA FA	FA		PS
12			IA FA	IA FA	FA	PS
13				IA FA	IA FA	
14				IA FA	IA FA	
15				FA		IA
16					FA	IA
17				FA	FA	IA
18	IA		IA	IA FA		
19			IA	IA	FA	

IA - Questions in Instructional Assessment
FA - Questions in Follow-up Assessment
PS - Practice Sheets

Classification by Question

Target Concepts and Abilities	Understand the vocabulary used in word problems	Use mental representations	Pay attention to the details of language and symbolization	Translate vocabulary into mathematical statements and vice versa	Understand the concept of variable by introducing place-holders	Recognize necessary and sufficient information
Question Number or Practice Sheet Number 20			IA	IA FA		FA
21			IA	IA	IA	
22			IA	IA	IA	
23				IA		IA

IA - Questions in Instructional Assessment
FA - Questions in Follow-up Assessment
PS - Practice Sheets

Instructional Assessment

The questions in both the Instructional and Follow-up Assessment Instruments for this unit have different formats appropriate to the nature of the concepts addressed in the questions.

It is necessary to introduce the students to these formats since they may have limited experience with different types of responses.

If there is not sufficient time to review examples of the different questions types, review the directions below with the students before they work the *Algebridge*™ exercises.

Directions

There are three types of questions in this exercise.

One type asks you to choose the answer to the problem and mark the corresponding letter, A, B, C, or D, on your answer sheet.

The second type gives you several choices related to a specific question. You must answer **yes** or **no** for each of these choices.

The third type asks you to write your own answer to the problem.

Focus of Questions 1–14

Questions 1–7 assess students' understanding of the phrase "times as many as." Similarly, questions 8–14 deal with the phrases "more than" and "is more than." Understanding such terms and being able to use them correctly is important to students' success in solving problems.

The exercises in these first two groups highlight the relationship between the language used in word problems and common student errors. In these exercises students grapple with the subtleties of language so that they can fully understand the differences between phrases with similar wording, such as "more than" and "is more than."

The questions in both of these groups progress from a simple interpretation of the mathematical statements to the construction of an equation. These first two groups of questions can serve as procedural models for students and teachers to follow as students work through word problems that involve similar terminology.

Questions 1–2

Questions 1–7 refer to the following statement.

Sharon has 4 times as many records as tapes.

1. What does Sharon have more of?
   (A) Records
   (B) Tapes

2. 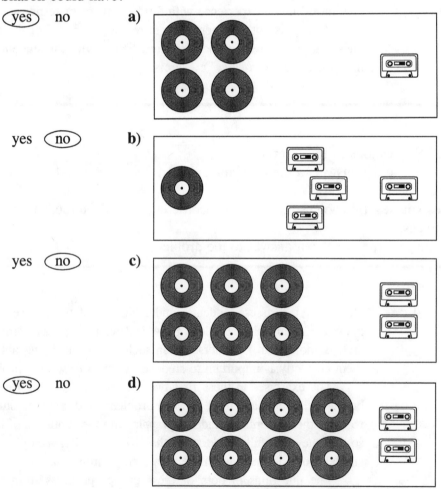 is one of Sharon's records.

   is one of Sharon's tapes.

   Consider each of the following. Does the picture represent the number of records and tapes that Sharon could have?

   yes no   a)

   yes no   b)

   yes no   c)

   yes no   d)

Analysis of Questions 1–2   Question 1 is the simplest assessment of students' understanding of what the phrase "times as many as" means. It asks for *qualitative* understanding of the relationship between the number of records and the number of tapes.

The most common error is to misinterpret the statement and assume that the number of tapes is more than the number of records.

Question 2 introduces a pictorial representation of the relationship between the quantities presented in the statement "Sharon has 4 times as many records as tapes." Some students may find it helpful to draw pictures such as these to illustrate mathematical relationships.

Guiding Class Discussion

The practice sheets and suggested activities in the Follow-up Instruction section under the concepts **Understand the vocabulary used in word problems** and **Use mental representations** may be helpful for those students having difficulty or needing more practice.

Practice Sheets 1–4 could be particularly useful at this point. For sheets 1 and 3, diagrams that represent problem situations are given. The student must label each diagram to correspond to the information given. For sheets 2 and 4, only the information is given. The student must produce a diagram to represent each situation.

Practice Sheet 5 probes for deeper understanding of the concepts by asking students to write a problem that could be represented by the given diagram.

Focus of Questions 3–5

Questions 3 and 4 assess *quantitative* understanding of the relationship between the number of records and the number of tapes. For those students who understand the statement "Sharon has 4 times as many records as tapes," question 5 presents different ways of saying the same thing.

Questions 3–5

Sharon has 4 times as many records as tapes.

3. Suppose that Sharon has 12 tapes. How many records does she have?

    (A)   3
    (B)   4
    (C)   16
    (D)   48

4. Consider each of the following. Do the numbers represent numbers of records and tapes that Sharon could have?

			Number of Records	Number of Tapes
yes	(no)	a)	3	12
(yes)	no	b)	12	3
yes	(no)	c)	12	8
(yes)	no	d)	20	5

121

5. Consider each of the following. Is it another way to express this sentence?

"There are 4 times as many records as tapes."

(yes)	no	**a)**	For every tape, there are 4 records.
yes	(no)	**b)**	There are 4 more records than tapes.
yes	(no)	**c)**	There are as many tapes as there are records.
(yes)	no	**d)**	The number of tapes multiplied by 4 gives you the number of records.
yes	(no)	**e)**	The number of tapes increased by 4 gives you the number of records.
(yes)	no	**f)**	4 times the number of tapes is equal to the number of records.

Analysis of Questions 3–5

In question 3, students will likely pick 4 (choice B) if they focus only on the number 4 only in the statement. Students will likely pick 3 (choice A) if they interpret the relationship to be the reverse of that intended. Students who pick 16 (choice C) may have added the numbers 4 and 12 together, interpreting the problem as "Sharon had 4 more records than tapes."

In question 5 incorrect phrasings of the statement may appear to be correct. Some phrasings use colloquial English while others use the kind of formal language used in mathematics classes.

Guiding Class Discussion

The practice sheets and suggested activities in the Follow-up Instruction section under the concepts **Understanding the vocabulary used in word problems**, **Use mental representations**, and **Pay attention to the details of language and symbolization** may be helpful for those students having difficulty or needing more practice.

In particular, **practice sheets 6, 7, and 9** focus on the use of tables to represent relationships between quantities. For these sheets, students complete or generate number pairs that satisfy the conditions of the given relationships. In addition, on practice sheet 7, the student must label the columns in the tables.

Focus of Question 6

Question 6 illustrates the first step in helping students translate from words to equations. The question uses a combination of words and symbols.

Question 6

6. Which of the following equations has the same meaning as this sentence?

"There are 4 times as many records as tapes."

(A) 4 + (number of records) = (number of tapes)
(B) 4 × (number of records) = (number of tapes)
(C) 4 + (number of tapes) = (number of records)
(D) 4 × (number of tapes) = (number of records)

Analysis of Question 6

Choice B is intended to diagnose the common student error of using a left- to-right word-order-match procedure to translate from the word problem to the equation. This choice gives an example of when a word-order match *cannot* be used to construct an equation for a word problem.

If students incorrectly answer this question but correctly respond to questions 3 and 4, encourage them to use questions 3 and 4 to help answer question 6.

Guiding Class Discussion

The practice sheets and suggested activities in the Follow-up Instruction section under the concepts **Pay attention to the details of language and symbolization** and **Translate vocabulary into mathematical statements and vice versa** may be helpful for those students having difficulty or needing more practice.

Focus of Question 7

In question 7, an additional level of abstraction is introduced by moving from the word representation for a variable used in question 6 to a placeholder representation for the variable. This question parallels question 6 and is designed to serve as a transition to algebra where letters are used to represent variables.

Question 7

7. Suppose that $\square$ stands for the number of records and $\triangle$ stands for the number of tapes.

Consider each of the following. Does the equation have the same meaning as this sentence?

"There are 4 times as many records as tapes."

(yes)	no	**a)**	$\square = 4 \times \triangle$
yes	(no)	**b)**	$4 \times \square = \triangle$
yes	(no)	**c)**	$4 + \square = \triangle$
yes	(no)	**d)**	$\square = 4 + \triangle$

Analysis of Question 7

Students may correctly answer question 6 but not question 7. Time should be spent showing students the relationship between these two problems and how the former can be used to help answer the latter. Teachers should ask students to compare the form of the answers for questions 6 and 7. Students should conclude that it is much shorter and more efficient to write mathematical statements in the form used in question 7.

Guiding Class Discussion

The practice sheets and suggested activities in the Follow-up Instruction section under the concepts **Translate vocabulary into mathematical statements and vice versa** and **Understand the concept of variable by introducing placeholders** may be helpful for those students having difficulty or needing more practice.

Focus of Questions 8–14

Questions 8–14, which deal with the phrases "more than" and "is more than," are parallel to the previous group of questions. Errors that the students are likely to make in answering these questions are similar to those described for questions 1–7. For those questions in this group that focus on concepts in a slightly different manner than the corresponding questions in the previous family, additional discussion is included. Sample classroom activities for this group of questions could parallel those suggested for the previous family with modifications appropriate to the different terms being used.

Questions 8–10

Questions 8–14 refer to the following statement.

> Last week at work, Carlos earned 7 dollars more than Tyrone.

8. Who earned more money?
   (A)   Carlos
   (B)   Tyrone

9. Suppose Tyrone earned 27 dollars. How much did Carlos earn?
   (A)   7 dollars
   (B)   20 dollars
   (C)   34 dollars
   (D)   189 dollars

10. Consider each of the following. Do the numbers represent amounts of money that Carlos and Tyrone could have earned last week?

		Number of Dollars That Carlos Earned	**Number of Dollars That Tyrone Earned**
yes (no)	a)	2	9
yes (no)	b)	7	1
(yes) no	c)	8	1
yes (no)	d)	14	2
yes (no)	e)	23	30
(yes) no	f)	30	23

Guiding Class Discussion

If students have difficulties with questions 8–10, a diagram showing the difference between the amounts that Carlos and Tyrone earned may be helpful. A figure can be drawn to show the relationship between the amount Carlos earned and the amount Tyrone earned. For example, if Carlos earned $15 and Tyrone earned $8, the following figure could be used to represent the relationship.

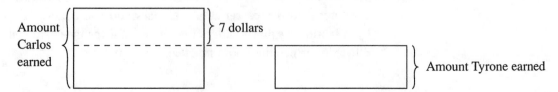

For the case when Carlos earned $30 and Tyrone earned $23, the figure could be

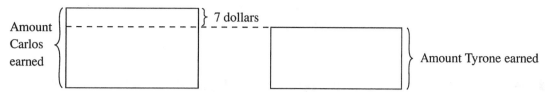

The teacher may wish to draw the students' attention to the difference in the diagrams for the two situations and discuss what portion of Carlos' earnings the $7 represents in each case.

Manipulatives may be useful in representing the relationship between the amount that Carlos earned and the amount that Tyrone earned.

As an extension of these activities, parallel activities using the phrases "earned 7 dollars less than" and "earned 7 times more than" will give the students additional practice with common phrases used in mathematics.

The practice sheets and suggested activities in the Follow-up Instruction section under the concept **Understand the vocabulary used in word problems** may be helpful for those students having difficulty or needing more practice.

Focus of Question 11

Question 11 addresses several common misinterpretations of the phrase "earned 7 dollars more than."

Question 11

11. Consider each of the following. Is it another way to express this sentence?

"Carlos earned 7 dollars more than Tyrone."

yes ~~no~~   **a)**   For every dollar Tyrone earned, Carlos earned 7 dollars.

~~yes~~ no   **b)**   The amount that Tyrone earned plus 7 dollars equals the amount that Carlos earned.

~~yes~~ no   **c)**   Tyrone earned 7 dollars less than Carlos.

yes ~~no~~   **d)**   Carlos earned 7 times as much as Tyrone.

~~yes~~ no   **e)**   The amount Carlos earned was 7 dollars greater than the amount Tyrone earned.

yes ~~no~~   **f)**   Carlos earned more than Tyrone.

Analysis of Question 11

Some students may answer **no** to 11c. Although 11c is mathematically equivalent to the problem statement, the real-world meaning is indeed different. ("More" and "less" do have different meanings.) A student should be given credit for answering **no** to 11c if the student can demonstrate that he or she recognizes that 11c is mathematically equivalent to the problem statement but interpreted the question to be asking also for an equivalence in the meaning of the language itself.

Note that 11f is a correct statement, but it is not an *equivalent* phrasing of the statement "Carlos earned 7 dollars more than Tyrone." Students who answer 11f incorrectly may not be paying attention to the details of the question or they may not understand that an *equivalent* phrasing is required.

Guiding Class Discussion

The practice sheets and suggested activities in the Follow-up Instruction section under the concept **Pay attention to the details of language and symbolization** may be helpful for those students having difficulty or who need more practice.

Focus of Question 12

As in question 6, question 12 illustrates the first step in translation from words to symbols. The question uses words and symbols such as > and =.

Question 12

12. Consider each of the following. Does the statement have the same meaning as this sentence?

"The amount Carlos earned is 7 dollars more than the amount Tyrone earned."

yes	(no)	**a)**	(amount Carlos earned) = 7 dollars > (amount Tyrone earned)
yes	(no)	**b)**	(amount Carlos earned) + 7 dollars > (amount Tyrone earned)
yes	(no)	**c)**	(amount Carlos earned) > (amount Tyrone earned)
yes	(no)	**d)**	(amount Tyrone earned) = (amount Carlos earned) + 7 dollars
(yes)	no	**e)**	(amount Carlos earned) = (amount Tyrone earned) + 7 dollars
(yes)	no	**f)**	(amount Tyrone earned) = (amount Carlos earned) − 7 dollars

---

Analysis of Question 12

Question 12a is intended to diagnose the common student error of using a left-to-right, word-order-match procedure to translate from the word problem to the equation. In this case, students who treat "more than" and "is greater than" as equivalent phrases *cannot* use this procedure to construct the equation. The teacher may want to contrast this use of "greater than" with the use in 11e.

Note that 12b and 12c are accurate statements, but are not equivalent to "the amount Carlos earned is 7 dollars more than the amount Tyrone earned."

Question 12f introduces subtraction as an equivalent way of representing an addition statement. Many students may correctly answer 12e but not 12f. It may be helpful to alert students to this equivalence.

Guiding Class Discussion

To illustrate the concepts of equality and inequality, the teacher may use a balance scale (or draw a picture of a balance scale if one is unavailable), using the balance or imbalance between the left- and right-hand sides to illustrate the concepts. Using the phrases "is equal to," "is greater than," and "is less than" to describe the relationship illustrated by the scale would help the students associate the appropriate terminology with the concepts of equality and inequality.

The figure below is one possibility for representing the phrasing used in question 12f,

(amount Tyrone earned) = (amount Carlos earned) − 7 dollars.

Covering the section of the figure that represents the 7 dollars may help the students see the relationship more clearly. By comparing this figure with the similar one drawn for questions 9 and 10, the student may be able to see the equivalence of 12e and 12f.

As in questions 9 and 10, diagrams representing several amounts for Carlos and Tyrone should be presented.

The practice sheets and suggested activities in the Follow-up Instruction section under the concepts **Pay attention to the details of language and symbolization** and **Translate vocabulary into mathematical statements and vice versa** may be helpful for those students having difficulty or needing more practice.

Focus of Questions 13–14

Questions 13 and 14 create a contrast between the phrases "7 dollars more than" and "more than."

Question 13

13. Suppose that □ stands for the amount Carlos earned and △ stands for the amount Tyrone earned.

Write an equation that means

"Carlos earned 7 dollars more than Tyrone."

Answer: *□ = △ + 7 (or any equivalent form)*

Analysis of Question 13

In *constructing* an equation, students are more likely to exhibit misconceptions or erroneous mental representations than if they are required only to *recognize* correct equations. Furthermore, by actively constructing their own equations, students are more likely to remember how to decipher the language in word problems. This activity thus parallels what students must do when solving word problems on their own since they will not be asked to recognize equations, but rather to construct equations from the statements made in a word problem.

Students sometimes confuse the concept of a variable with the concept of a label for a quantity. They need to see that there is a difference between "dollars" and "the number of dollars that Carlos (or Tyrone) has."

Although there is only one question in the Instructional Assessment that asks students to construct their own equation, there are many such questions in the Follow-up Assessment, highlighting the importance of this activity.

Guiding Class Discussion

Arithmetic word problems can be used to help students gain skill in representing the mathematical relationships presented in algebra word problems. By having the answer to the arithmetic word problem, students can focus on the method used to reach a solution rather than on the solution itself. For example, the arithmetic problem might be the following:

> At the movies, an adult ticket costs $4.50, and a student ticket costs $2.50. Carlos went to the movies with some friends. He paid for 2 adult tickets and 3 student tickets. How much did Carlos spend on tickets?

For this purpose, we would change the last line to read "Carlos spent a total of $16.50" and add "Write a number sentence that shows how to calculate how much Carlos spent on the tickets."

Arithmetic texts can be used as a source of problems for this activity.

The practice sheets and suggested activities in the Follow-up Instruction section under the concepts **Translate vocabulary into mathematical statements and vice versa** and **Understand the concept of variable by introducing placeholders** may be helpful for those students having difficulty or needing more practice.

Practice sheets 8 and 10, in particular, focus on the vocabulary-mathematical statement concept. For practice sheet 8, students complete tables that are labelled with equations that represent given relationships. For practice sheet 10, students must write a statement that could be represented by the given table.

Question 14

14. Suppose that $\square$ stands for the amount Carlos earned and $\triangle$ stands for the amount Tyrone earned.

Consider each of the following. Does it mean the same as this sentence?

"Carlos earned more than Tyrone."

(yes) no	**a)**	$\square > \triangle$	
yes (no)	**b)**	$\square = \triangle$	
yes (no)	**c)**	$\triangle > \square$	
yes (no)	**d)**	$\square + \triangle$	

Guiding Class Discussion

To help determine if students understand the subtle distinctions between "7 dollars more than" in question 13 and "more than" in question 14, the teacher might ask such questions as the following:

1. Which statement describes the general relationship between the amounts earned by each person?

2. Which statement is more specific (or gives more information) about the amount earned by each person?

3. Which statement tells which person earned more money?

4. If you knew how much money Tyrone earned, which statement could be used to determine exactly how much Carlos earned?

To emphasize the distinction between the two phrases, the teacher could complete the chart below with the students.

Statement	Amount Tyrone Earned	Amount Carlos Earned
Carlos earned 7 dollars more than Tyrone. $\square = \triangle + 7$	$5	
Carlos earned more than Tyrone. $\square > \triangle$	$5	

The teacher should ensure that the students realize that $\square > \triangle$ really means "the amount Carlos earned is more than the amount Tyrone earned."

The practice sheets and suggested activities in the Follow-up Instruction section under the concepts **Translate vocabulary into mathematical statements and vice versa** and **Understand the concept of variable by introducing placeholders** may be helpful for those students having difficulty or needing more practice.

Focus of Questions 15–17

The purpose of questions 15–17 is to assess whether students are able to recognize relevant information necessary to solve problems.

Questions 15–17

15. Greg is paid 10 cents for each newspaper he delivers.

Consider each of the following. Is the information necessary for figuring out how much money Greg makes each week delivering newspapers?

yes (no) **a)** The selling price of each newspaper
yes (no) **b)** The number of pages in the newspaper
yes (no) **c)** The number of people who live in Greg's town
(yes) no **d)** The number of papers Greg delivers each week

16. In the following problem, cross out all information that will NOT help you figure out how far Mei-Ying traveled on her bicycle trip. (Note: Do not solve the problem!)

> Last Saturday Mei-Ying went to her aunt's house on her bicycle. She left home at 9:00 a.m. and rode until 9:30 a.m. At 9:30, she stopped for a break and ~~bought a bottle of juice for 75 cents~~. After a 15-minute break, Mei-Ying continued on her bicycle, arriving at her aunt's house at 10:30 a.m. Her average speed for the trip was 8 miles per hour. How many miles did Mei-Ying travel in all?

17. Nicole was paid $4 per hour to paint her neighbor's garage. What else do you need to know to figure out how much she made painting the garage?

Answer: *The number of hours Nicole spent painting the garage.*

Analysis of Questions 15–17

If students answer 15d correctly but answer 15a–15c incorrectly, they should explain why the information is or is not necessary to solve the problem. By challenging the students to provide reasons for their answers, the teacher can help develop analytical problem-solving and communication skills.

Question 16 asks the students to identify extraneous information by crossing it out. Although the students are not often required to deal with extraneous information in textbook problems, understanding this aspect of a problem may lead to more refined problem-solving techniques. The goal is for the students to be able to comprehend the information presented in light of what they are being asked to find in a given problem.

Question 17 also builds problem-solving skills by encouraging students to think about what additional information they would need to solve a problem. Responses to this question may vary. Students may write "how long" rather than "how many hours." They may also include additional information such as "the amount she spent on supplies for painting her garage" since Nicole may have to subtract this from her earnings. A student who provides such an answer may be thinking in terms of net earnings rather than gross earnings. Teachers should give credit for answers that students can defend as valid.

Guiding Class Discussion

The practice sheets and suggested activities in the Follow-up Instruction section under the concept **Recognize necessary and sufficient information** may be helpful for those students having difficulty or needing more practice.

In particular, **practice sheet 11**, like question 17, asks students to determine what additional information is requireed to solve the given problems. **Practice sheet 12** parallels question 16, asking students to cross out any information that is not needed to solve the given problems.

Focus of Question 18

Question 18 assesses the students' ability to recognize appropriate mathematical equations that can represent the relationship "half as many as."

Question 18

18. The Maple High School soccer team won half as many games as the basketball team won. The basketball team won 12 games.

Consider each of the following. Does the equation show how many games the soccer team won?

yes (no)   **a)**   Number of soccer games won $= 12 \times 2$

(yes) no   **b)**   Number of soccer games won $= \dfrac{1}{2} \times 12$

yes (no)   **c)**   Number of soccer games won $= \dfrac{1}{2} + 12$

(yes) no   **d)**   Number of soccer games won $= \dfrac{12}{2}$

yes (no)   **e)**   Number of soccer games won $= 12 \div \dfrac{1}{2}$

Analysis of Question 18

In addition to possible confusion caused by interchanging the two teams, students may not recognize the equivalence of multiplying by $\frac{1}{2}$ and dividing by 2. This equivalence is used extensively in algebraic manipulations with variables and should be emphasized in the classroom. If many students correctly answer 18b but incorrectly answer 18d (or vice versa), time should be spent explaining the equivalence.

Guiding Class Discussion

If students have difficulty with this question, manipulatives could be used to help them get a concrete idea of the equivalence of multiplying by $\frac{1}{2}$ and dividing by 2. For example, start with 10 toothpicks and model the problem $10 \div 2$ by separating 10 toothpicks into groups of 2; the result is 5. The problem $10 \times \frac{1}{2}$ yields the same result as $10 \div 2$ since $10 \times \frac{1}{2}$ is the same as $\frac{1}{2} \times 10$ and taking $\frac{1}{2}$ of a group of 10 toothpicks yields a result of 5. It also may be appropriate at this point for the students to use manipulatives to investigate what it means to *divide* by $\frac{1}{2}$. For example, the problem $10 \div \frac{1}{2}$ could be modeled by separating 10 toothpicks into groups of $\frac{1}{2}$. This is accomplished by breaking each toothpick in half, yielding 20 halves.

The practice sheets and suggested activities in the Follow-up Instruction section under the concepts **Understand the vocabulary used in word problems**, **Pay attention to the details of language and symbolization**, and **Translate vocabulary into mathematical statements and vice versa** may be helpful for those students having difficulty or needing more practice.

Focus of Questions 19–20

Questions 19 and 20 involve translating information in a word problem into a numerical equation.

Questions 19–20

19. Sasha is 13 years older than Ted.
Ted is 8 years old.

Which of the following shows how to calculate Sasha's age?

- (A) Sasha's age = 13 + 8
- (B) Sasha's age = 13 − 8
- (C) Sasha's age = 13 × 8

20. Kevin is 5 years older than twice Barbara's age.
Barbara is 7 years old.

Which of the following shows how to calculate Kevin's age?

- (A) Kevin's age = (5 + 2) × 7
- (B) Kevin's age = 5 > 2 × 7
- (C) Kevin's age = 5 + (2 × 7)
- (D) Kevin's age = (5 × 2) + 7

Analysis of Question 20

Question 20 assesses whether the student understands the proper order of operations in an equation that involves both multiplication and addition. That is, do students realize that you first multiply and *then* add in this equation? The number order in the equations is the same (5,2,7) so that the student is forced to decide which *operation* order is correct.

Note that choice (B), although not mathematically correct, is a direct translation of the words in the problem to symbols, with > representing "older than." If students choose this answer, they should be asked to explain why they chose it and helped to understand why it is not correct.

Guiding Class Discussion

To help students understand the differences between answers such as (5 + 2) × 7 and (2 × 7) + 5, the teacher should follow the pattern of activities depicted for the two families of questions in the beginning of this assessment and apply it to this problem. For example, if students have difficulties understanding that 2 × 7 + 5 is the correct answer, they could be encouraged to ask themselves who is older, Kevin or Barbara. Then they could draw a diagram, such as the one shown below, and write the equation using words, as in 2 × (Barbara's age) + 5 = (Kevin's age).

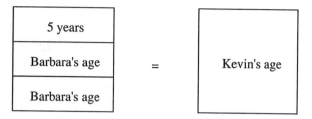

The practice sheets and suggested activities in the Follow-up Instruction section under the concepts **Pay attention to the details of language and symbolization** and **Translate vocabulary into mathematical statements and vice versa** may be helpful for those students having difficulty or needing more practice.

Focus of Questions 21–22

Questions 21 and 22 are intended to show students that the order of the symbolic representation of a word problem is not necessarily the same as the word order in the statement given in the word problem. Question 21 involves a single operation while question 22 involves two operations.

Questions 21–22

21. Let $\square$ stand for some number. Which of the following stands for

"6 less than the number"?

(A) $\square - 6$
(B) $6 - \square$
(C) $\square < 6$
(D) $6 < \square$

22. Let $\triangle$ stand for some number. Which of the following stands for

"8 less than twice the number"?

(A) $8 - (2 \times \triangle)$
(B) $(2 \times \triangle) - 8$
(C) $8 < 2 \times \triangle$
(D) $2 - (8 \times \triangle)$

Analysis of Questions 21–22

In question 22, many students might choose (A) or (C) as the correct answer since they are the most literal left-to-right, word-order-match translations.

Guiding Class Discussion

It may be helpful to conduct a class discussion about the important distinction between questions 20 and 22 to highlight the idea that although addition is commutative, subtraction is not.

- In question 20, the phrase "5 years older than" implies addition, and therefore the computation order is not important in the sense that $5 + (2 \times 7)$ and $(2 \times 7) + 5$ are both correct.

- In question 22, because of the phrase "8 less than ...," the problem involves subtraction. Because subtraction is not commutative, the order in which the mathematical expression is written is very important. Students may need to substitute a value of $\square$ to see why $2 \times \square - 8$ and $8 - 2 \times \square$ are *not* both correct representations of the statement.

Students also need to see that the presence of one additional word in a phrase can mean the use of a different symbol. For example, the phrase "less than" signifies the use of $-$ while the phrase "is less than" signifies the use of $<$. To help students determine when to use the appropriate symbol, ask them to write a symbolic statement for several phrases such as "10 is less than 15" and "10 less than 15." Include a variable in some of the phrases as well as an example using "less" ("22 less 12") to underscore the need for paying attention to the words used in a problem statement. Students should contrast the various symbolic statements and be able to say and write appropriate phrases if given the symbolic form.

The practice sheets and suggested activities in the Follow-up Instruction section under the concepts **Pay attention to the details of language and symbolization**, **Translate vocabulary into mathematical statements and vice versa**, and **Understand the concept of variable by introducing placeholders** may be helpful for those students having difficulty or needing more practice.

Focus of Question 23

While solving word problems, students must work from the words to the equation; they are rarely asked to perform the inverse translation. In this question, however, students are given an equation and asked to determine if a given situation fits the equation.

Question 23

23. Consider each of the following. Can the equation

$$5 \times 3 + 2 = 17$$

represent the statement?

(yes) no	**a)**	Five tickets at \$3 each plus a \$2 ticket cost \$17.
(yes) no	**b)**	Three \$5 lunches and a \$2 tip come to \$17.
yes (no)	**c)**	Akira rode his bicycle 5 miles in 3 weeks and 2 weeks later he rode 17 miles.

yes (no)   **d)**   Jennifer walked 5 miles at 3 miles per hour and then walked 2 more miles for a total of 17 miles.

(yes) no   **e)**   Susana is 17, which is 2 more than 5 times the age of her three-year-old sister.

Analysis of Question 23

The process of working backward, that is, starting with an equation and creating situations that fit the equation, is a procedure that is not often used. But the task of writing a "story line" to match an equation involves some higher-level cognitive processes that are important for algebraic problem solving. Students who answer 23c or 23d incorrectly may be confused about rates or may not recognize irrelevant information.

Guiding Class Discussion

If students have difficulties with question 23, the teacher can ask them to construct an equation for each of the statements given and then to compare their equations with the one given in the question.

An additional activity would be to have the students create numerical equations of varying difficulty, such as $3 + 2 = 5$, $7(3) + 2 = 23$, and $5(4) - 2(4) = 12$, and then write several word problems that could be represented by each of these equations.

The practice sheets and suggested activities in the Follow-up Instruction section under the concepts **Translate vocabulary into mathematical statements and vice versa** and **Recognize necessary and sufficient information** may be helpful for those students having difficulty or needing more practice.

Practice sheets 5 and 10 also focus on the "working backward" process by asking students to write problems that could be represented by the given diagram or table, respectively.

Follow-up Instruction

After the Instructional Assessment has been used to pinpoint students' conceptual weaknesses, the teacher may wish to use the following suggested activities and practice sheets to help correct those misconceptions. Because of the interrelated nature of this topic, each practice sheet involves a number of concepts. Practice sheets 1–5 involve using diagrams to understand the information presented in a problem while practice sheets 6–10 involve using tables to understand information. Sheets 11 and 12 provide practice in identifying what information is necessary to solve a problem.

Understand the vocabulary used in word problems

Practice sheets 1–10 as well as Follow-up Assessment questions 1–3 and 10 address this concept.

Use mental representations

Activity. If students have difficulties with the first three questions in the Instructional Assessment, the teacher should consider drawing a diagram showing the relationship between the number of records and the number of tapes that Sharon has. For example, the following figure is one way to show this relationship.

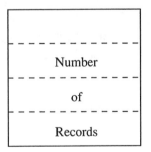

It may also be helpful to develop systematically a table (not unlike the one in question 4) to show different possibilities for the number of records and the number of tapes and have students check that the table is consistent with the problem statement and the diagram above.

For further practice students could draw diagrams to represent similar situations they make up or the teacher makes up. If students make up their own situations, they can work with a partner and take turns drawing and interpreting diagrams.

A parallel activity using phrases such as "Sharon has 4 more records than tapes" and "Sharon has 4 less records than tapes" would create contrasts for the students and therefore deepen their understanding.

Activity. If students have difficulties with question 2 or have problems differentiating between the records and tapes in the figures, it may be helpful to use manipulatives (*e.g.*, 4 records and 2 tapes or 5 erasers and 3 pieces of chalk) to represent different possibilities for the number of records and tapes. Then the students could draw a diagram that shows the mathematical relationship between the number of objects. Ask the students to talk about the difference between a photograph of the objects themselves and the diagram that they drew.

Practice sheets 1–5 as well as Follow-up Assessment question 9 also address this concept.

Pay attention to the details of language and symbolization

Activity. If the students have difficulty with question 5, the teacher may present a statement such as "Ron has 3 times as many pencils as Sam has" and ask students to generate a set of phrases that could be used to express the relationship between the number of pencils Ron has and the number of pencils Sam has. The teacher may need to give some students more guidance by asking them to consider the following questions:

1. Who has more pencils, Ron or Sam?
2. If Sam has 1 pencil, how many pencils does Ron have?
3. For every pencil Sam has, how many pencils does Ron have?
4. If you know the number of pencils Sam has, how can you determine the number of pencils Ron has?

The phrases generated by the students could then be used to write sentences that say the same thing as "Sharon has 4 times as many records as tapes." The students can then draw a figure and check to see that each of the sentences that they generated fits the figure.

These activities could be extended to fit similar terminology in other contexts. The list of terms associated with this unit can be used as a reference.

Practice sheets 1–10 as well as Follow-up Assessment questions 4, 5, and 10–12 also address this concept.

Translate vocabulary into mathematical statements and vice versa

Practice sheets 1–10 as well as Follow-up Assessment questions 5–7, 10–15, 17, 18, and 20 address this concept.

Understand the concept of variable by introducing placeholders

Activity. If students have difficulties with question 7, a set of numbers chosen from the chart in question 4 can be written inside the placeholders representing the number of records and tapes. The students then need to check whether the numbers satisfy each of the mathematical relationships shown.

The value of paying attention to detail in these types of problems will become evident to students if they mix up the numbers in the respective placeholders. The teacher may want to check for this type of error.

This activity helps to illustrate how symbols are used to represent mathematical relationships. It also shows consistency between the mathematical and verbal statements describing the number of records and tapes.

Using the list of terms in this unit for suggestions, the teacher could give the students some other verbal statement describing a mathematical relationship between two or more quantities. The students' task would be to write a mathematical equation with placeholders that represents the relationship between the quantities and then write numbers in the placeholders to verify that the equation is correct.

Practice sheet 5 as well as Follow-up Assessment questions 6, 7, 12–14, 16, 17, and 19 also address this concept.

Recognize necessary and sufficient information

Practice sheets 11 and 12 as well as Follow-up Assessment questions 8 and 20 address this concept.

Bibliography of Resource Materials

Crandall, Jody, et al. *English Skills for Algebra.* Center for Applied Linguistics, Washington, D.C., Englewood Cliffs, N.J.: Prentice-Hall, 1987.

Hansen, Merrily P., and Maryann Marropodi. *Practicing Problem-Solving.* Garber Group, Random House, 1982.

Leitzel, Joan R., and Franklin D. Demana. *Getting Ready for Algebra.* Lexington, MA: D.C. Heath, 1988. (A software package is available in addition to the text.)

Lochhead, Jack, and Jose Mestre. "From Words to Algebra: Mending Misconceptions." *1988 Yearbook of the National Council of Teachers of Mathematics*: 127–135.

Answers

Answers to Instructional Assessment

1. A

2. a) yes
b) no
c) no
d) yes

3. D

4. a) no
b) yes
c) no
d) yes

5. a) yes
b) no
c) no
d) yes
e) no
f) yes

6. D

7. a) yes
b) no
c) no
d) no

8. A

9. C

10. a) no
b) no
c) yes
d) no
e) no
f) yes

11. a) no
b) yes
c) yes
d) no
e) yes
f) no

12. a) no
b) no
c) no
d) no
e) yes
f) yes

13. $\square = \triangle + 7$ or any equivalent form

14. a) yes
b) no
c) no
d) no

15. a) no
b) no
c) no
d) yes

16. The information "bought a bottle of juice for 75 cents" is not needed.

17. The number of hours Nicole spent painting the garage

18. a) no
b) yes
c) no
d) yes
e) no

19. A

20. C

21. A

22. B

23. a) yes
b) yes
c) no
d) no
e) yes

Answers to Practice Sheets

Practice sheet 1
1. Mother, Sandra
2. A, B
3. Sylvia, Darlene
4. Tom, Neil
Practice sheet 2
Answers will vary.
1.

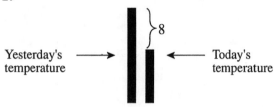

Yesterday's temperature → ← Today's temperature

2.

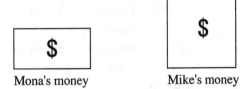

Mona's money   Mike's money

3.

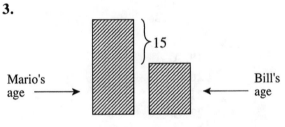

Mario's age →   ← Bill's age

4.

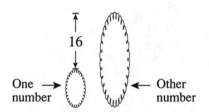

One number →   ← Other number

5.

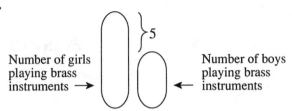

Number of girls playing brass instruments →   ← Number of boys playing brass instruments

Practice sheet 3
1. *A, B*
2. Last, This
3. Rob, Carmella
4. record, cassette tape
5. 8 more than a number, the number
Practice sheet 4
Answers will vary.
1.

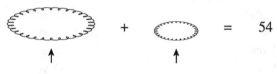

First number   Second number

2.

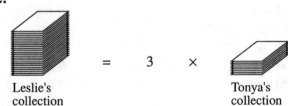

Leslie's collection   =   3   ×   Tonya's collection

3.

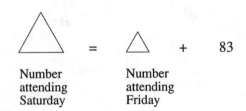

Number attending Saturday   =   +   83

Number attending Friday

4.

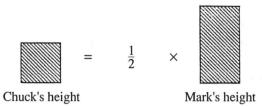

Chuck's height   Mark's height

5.

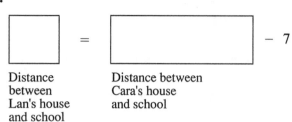

Distance between Lan's house and school

Distance between Cara's house and school

Practice sheet 5

Answers will vary.

1. The number of girls in a group is 12 more than the number of boys. If there are 15 boys, how many girls are there in the group?

2. The number of red marbles is 9 times the number of blue marbles. If there are 14 blue marbles, how many red marbles are there?

3. If one-half of a number subtracted from the number gives a result of 70, what is the number?

Practice sheet 6

1. 1, 5

3, 7

10, 14

For example 17, 21

For example 30, 34

2. 14, 19

9, 14

For example 12, 17

For example 21, 26

3. $124, $62

$780, $390

For example $320, $160

For example $90, $45

4. $15, $9

$80, $48

$45, $27

For example $20, $12

For example $50, $30

Practice sheet 7

Answers will vary.

1.

Number A	Number B
5	1
10	2
15	3
20	4
25	5

2.

Maria's Age	Keith's Age
10	1
11	2
12	3
13	4
14	5

3.

Number of Records Akia Has	Number of Records Linda Has
1	3
5	15
6	18
10	30
14	42

4.

Number in Algebra	Number in Geometry
21	10
27	14
30	16
29	19
33	22

Practice sheet 8

1. $29 - 15 = 14$

$82 - 68 = 14,$

$79 - 65 = 14$

$154 - 140 = 14$

2. $18 = 13 + 5$

$71 = 13 + 58$

$90 = 13 + 77$

$211 = 13 + 198$

3. $5(8) = 40$

$5(27) = 135$

$5(16) = 80$

$5(43) = 215$

For example $5(60) = 300$

For example $5(24) = 120$

Practice sheet 9
Answers will vary.

1.
Lars' Time	Jan's Time
500	504
620	624
652	656
710	714
803	807

2.
Number of Current Members	Number of Original Members
24	3
56	7
72	9
112	14
176	22

3.
Price of Sundae	Price of Sandwich
$1.20	$ .40
$1.35	$ .45
$1.80	$ .60
$2.25	$ .75
$2.70	$ .90

4.
Tim's Age	Mrs. Mason's Age
4	27
6	29
11	34
14	37
21	44

Practice sheet 10
Answers will vary.
1. A number X is 12 times a number Y.
2. The distance from school to Rita's house is 3 miles more than the distance from school to Ned's house.
3. The difference between the number of hamburgers and hot dogs sold was 18. There were more hamburgers than hot dogs sold.

Practice sheet 11
1. The price of the stereo system
2. The distance Connie ran in any one of the five days
3. The total number of points scored by both teams *or* the number of points scored by the Main Street team
4. The length in inches on the map between Fernville and Oceantown
5. Answers will vary. For example, amount of profit on a medium pizza.

Practice sheet 12
1. The information "in 2 buses" is not needed.
2. The information "and its perimeter is 34" is not needed.
3. The information "gasoline costs $1.15 per gallon" is not needed.
4. The information "There are twice as many cloudy days in Town B as in Town A" is not needed.
5. All information is relevant.

Answers to Follow-up Assessment

1. A

2. C

3. a) no

  b) no

  c) yes

  d) yes

  e) no

4. a) no

  b) yes

  c) no

  d) yes

  e) no

5. a) no

  b) no

  c) no

  d) no

  e) yes

  f) yes

6. a) yes

  b) no

  c) no

  d) no

7. $\Box = \triangle + 4$

  (or any equivalent form)

8. How many cans of oil
  Lisa bought and how much
  each gallon of gas costs

9. For example,

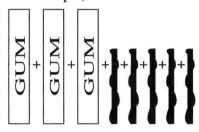

  $.50 $.50 $.50 $.35 $.35 $.35 $.35 $.35

10. a) no

  b) yes

  c) no

  d) yes

  e) no

11. C

12. a) yes

  b) yes

  c) no

  d) no

13. (amount Linda saved)
  = (amount Jiro saved)
  − $5.80

14. (amount Linda saved)
  = $10.25 − $5.80

15. (amount Linda saved)
  < (amount Jiro saved)

16. a) no

  b) no

  c) no

  d) no

  e) yes

17. $7 - 4 = \triangle$

  (or any equivalent form)

18. $(.35 \times 12) - 2.95 = 1.25$

19. $1.25 \times \Box = \triangle$

20. a) yes

  b) no

  c) no

  d) yes

  e) no

ALGE**BRIDGE**™

Attacking

Word Problems

Successfully

Attacking Word Problems Successfully

Introduction

This unit helps students develop sound techniques for attacking any word problem. To attack a problem successfully, the student must know what questions to ask about the problem in order to determine what each quantity means and how it relates to other quantities in the problem. Knowing how to attack a problem helps develop a sense of what to do with the numbers in a problem and lays a good foundation for developing useful problem-solving techniques.

The word problems presented in this unit include situations that are similar to those typically found in pre-algebra and algebra texts. Unlike the typical text, however, the focus in this unit is on the underlying concepts and abilities involved in understanding and interpreting the information given in such problems. While the intent of the unit is not to teach solution methods for specific types of word problems (*i.e.*, age, coin, mixture, etc.), such problems can provide a starting point for learning how to attack any word problem. Note that some of the practice sheets include problems of a less routine nature which allow students to practice and extend their problem-solving skills.

In the context of word problems that a student may encounter in pre-algebra or of situations from everyday life, this unit develops the student's ability to

– determine what is being asked;

– identify given and assumed information relevant to the problem's solution;

– use given information to set up a solution procedure;

– estimate and determine the reasonableness of a problem's answer;

– use diagrams, charts, or tables to represent/interpret information;

– recognize possible alternate solution methods for a problem.

Cross-References to
Pre-Algebra and Algebra
Textbook Topics

The concepts and abilities addressed in this unit are not often treated explicitly in pre-algebra or algebra textbooks. Word problems appear throughout the textbooks, however, and the use of this unit would be helpful when students are introduced to word problems.

This unit may also be helpful when studying the following topics that appear in many pre-algebra and algebra textbooks.

algebraic expressions   **variables**
open sentences   **word problems**
problem solving

List of Terms

The following is a list of words and phrases that are frequently used in word problems. The teacher may want to add other words and phrases found in word problems.

area	**double**	**price**	**time and a half**
average	**equation**	**rate**	**triangle**
border	**expression**	**rectangle**	**unit**
cost	**fraction**	**region**	**word problem**
dimensions	**least**	**sales tax**	**years ago**
distance	**most**	**temperature**	**years from now**

Classification by Concept

Target Concepts and Abilities	Questions in Instructional Assessment	Questions in Follow-up Assessment	Practice Sheets
Determine what is being asked	4, 8, 14, 26	4, 8, 14, 26	7
Identify given information relevant to a problem's solution	3, 10, 16, 17	3, 10, 16, 17	1, 5, 6, 7
Identify or use assumed information relevant to a problem's solution	5, 9	5, 9	6, 7
Use given information to set up a solution procedure	2, 7, 11, 12, 13, 15, 19, 20, 23, 25	2, 7, 11, 12, 13, 15, 19, 20, 23, 25	6, 7
Estimate and determine the reasonableness of a problem's solution	1, 18, 27	1, 18, 27	7
Represent given information by diagrams, charts, or tables	6, 21, 22, 24, 28	6, 21, 22, 24, 28	2, 3, 4, 7
Recognize possible alternate methods for a problem's solution	2, 7, 13, 20, 23, 29	2, 7, 13, 20, 23, 29	2, 6, 7

Classification by Question

Target Concepts and Abilities	Determine what is being asked	Identify relevant given information	Identify relevant assumed information	Set up a solution procedure	Determine the reasonableness of a solution	Represent information by diagrams, etc.	Recognize alternate methods for a solution
Question Number or Practice Sheet Number 1		PS			IA FA		
2				IA FA		PS	IA FA PS
3		IA FA				PS	
4	IA FA					PS	
5		PS	IA FA				
6		PS	PS	PS		IA FA	PS
7	PS	PS	PS	IA FA PS	PS	PS	IA FA PS
8	IA FA						
9			IA FA				
10		IA FA					
11				IA FA			
12				IA FA			
13				IA FA			IA FA
14	IA FA						
15				IA FA			
16		IA FA					
17		IA FA					

IA - Questions in Instructional Assessment
FA - Questions in Follow-up Assessment
PS - Practice Sheets

Classification by Question

Target Concepts and Abilities		Determine what is being asked	Identify relevant given information	Identify relevant assumed information	Set up a solution procedure	Determine the reasonableness of a solution	Represent information by diagrams, etc.	Recognize alternate methods for a solution
Question Number or Practice Sheet Number	18					IA FA		
	19				IA FA			
	20				IA FA			IA FA
	21						IA FA	
	22						IA FA	
	23				IA FA			IA FA
	24						IA FA	
	25				IA FA			
	26	IA FA						
	27					IA FA		
	28						IA FA	
	29							IA FA

IA - Questions in Instructional Assessment
FA - Questions in Follow-up Assessment
PS - Practice Sheets

Instructional Assessment

The questions in both the Instructional and Follow-up Assessment Instruments for this unit have different formats appropriate to the nature of the concepts addressed in the questions.

It is necessary to introduce students to these formats since they may have limited experience with different types of responses.

If there is not sufficient time to review examples of the different question types, review the directions below with the students before they work the *Algebridge*™ exercises.

Directions

There are three types of questions in this exercise.

One type asks you to choose the answer to the problem and mark the corresponding letter, A, B, C, or D, on your answer sheet.

The second type gives you several choices related to a specific question. You must answer **yes** or **no** for each of these choices.

The third type asks you to write your own answer to the problem.

Focus of Question 1

Question 1 assesses whether a student can establish the boundaries for a problem's answer. This question exposes the student to an indirect type of estimation in a real-world context.

Question 1

1. Shirts sell for $12 each. A sales tax of 5 percent must be paid for each shirt purchased. Answer each of the following questions.

(yes) no	**a)**	Sue has exactly $20. Can she buy one shirt?
yes (no)	**b)**	John has exactly $12. Can he buy one shirt?
(yes) no	**c)**	Tresha has exactly $24. Can she buy one shirt?
(yes) no	**d)**	Jeff has exactly $16. Can he buy one shirt?
yes (no)	**e)**	Sam has exactly $10. Can he buy one shirt?
(yes) no	**f)**	Dana has exactly $26. Can he buy *two* shirts?

Analysis of Question 1

Students who answer **yes** to 1b or 1e may not have understood that, because of the tax, more than $12 will be needed to buy the shirt. The teacher may want to mention that no calculations are necessary to answer 1b and 1e correctly.

Students who answer **no** to 1a, 1c, or 1d may not understand how to estimate 5 percent of 12. For a student to estimate the amount needed to buy one shirt, he or she should be able to determine that 5 percent of $12 is less than $1. Therefore, anyone who has $12 + $1 (or $13) will have enough money to buy one shirt. Students who answer **no** to 1f may have difficulty working with percents and purchasing multiple quantities in the same problem.

Guiding Class Discussion

Encourage the students to explain why each of the amounts is enough or not enough to purchase a shirt.

It should be pointed out to students that in real-world situations, the consumer is often interested in determining a whole number only slightly above the amount of money needed for a certain purchase. For example, to purchase 3 records at $6.79 each and pay a 6 percent sales tax on the records, one might estimate that $25 would cover the purchase. Responses such as $100, $500, or $1,000, although correct, are not reasonable.

For those students who are having difficulties with the ideas presented in this question, the practice sheets and suggested activities in the Follow-up Instruction section under the concept **Estimate and determine the reasonableness of a problem's solution** may be helpful.

Focus of Questions 2–3

Questions 2 and 3 assess whether a student can work with information presented in a problem and recognize that different methods can be used to solve the same problem. Processing information involves identifying explicitly stated relevant information as well as using outside knowledge to solve the problem. It also includes understanding *how* this information can be used to solve the problem.

Questions 2–3

2. Jay wants to buy a shirt that costs $15. The sales tax on the shirt is 6 percent. Consider each of the following. Does the statement *completely* describe a method that could be used to find the amount of money Jay needs to buy the shirt?

yes **(no)**   **a)** Multiply the price of the shirt by 0.06.

(yes) no   **b)** Multiply the price of the shirt by 1.06.

(yes) no   **c)** Multiply the price of the shirt by 0.06 and add that amount to the price of the shirt.

yes **(no)**   **d)** Add 0.06 to the price of the shirt.

3. All books in a certain discount store are sold at 20 percent less than their suggested retail price. Lynn wants to buy two different books from this store. Consider each of the following facts.

Would knowing the fact help you to determine how much money Lynn would need to pay for the books?

yes (no)	**a)**	Which of the two books costs more
yes (no)	**b)**	The number of pages in each book
(yes) no	**c)**	The suggested retail price of each book
yes (no)	**d)**	The amount of profit that the store makes on each book
(yes) no	**e)**	Amount of discount on each book = 0.20 × suggested retail price of that book
yes (no)	**f)**	The total number of books in the store
(yes) no	**g)**	80 percent of the suggested retail price = the discounted price of the book

Analysis of Questions 2–3

Students who answer **yes** to 2a may not understand that the sales tax must be *added* to the price of the goods being taxed. (They can be credited with a correct answer if they explain that 2a can be used to find the amount of tax, which is then used to find the total amount.) The solution method suggested in 2b may be especially difficult for students to recognize. Students will need to understand that 1.06 is equivalent to 106 percent, which can be expressed as 100 percent + 6 percent (where 100 percent is the cost of the shirt and 6 percent is the sales tax). A **no** answer to 2c combined with a **yes** answer to 2b, or vice versa, indicates that the student may not be aware that there are different ways to solve the problem. A **yes** answer to 2d indicates that the student could benefit from an explanation of how to find a percent of a number.

Students who answer **yes** to 3a, 3b, 3d, or 3f need to understand that this information is *not* relevant to the solution of the problem. A class discussion might help students see that, although this information might be useful, it would not be helpful in solving the problem. Encourage the students to explain their answers. A discussion of 3g should help students realize that 80 percent of a given price is the same as 20 percent less than the given price.

Guiding Class Discussion

For those students who are having difficulties or needing more practice with the ideas presented in these questions, the practice sheets and suggested activities under the concepts **Use given information to set up a solution procedure**, **Recognize possible alternate methods for a problem's solution**, and **Identify given information relevant to a problem's solution** in the Follow-up Instruction section may be helpful.

Focus of Questions 4–7   Questions 4–7 take the student through a group of questions in a logical problem-solving progression. One begins by determining the correct form of the answer (question 4), then identifies relevant information (question 5), and concludes by representing information and recognizing different ways to attack the problem (questions 6 and 7).

Questions 4–7

Questions 4–7 refer to the following story problem.

The temperature was 70° at 8 a.m. on a certain day. During the day the temperature went up 2° each hour. What was the temperature at 3 p.m. that day?

4. The answer will be a
    (A) temperature
    (B) time of day
    (C) number of days
    (D) number of hours

5. Which of the following do you need to know to find the temperature at 3 p.m. that day?
    (A) The day of the week
    (B) The temperature at noon
    (C) The number of hours between 8 a.m. and 3 p.m.
    (D) The number of hours in a day

6.

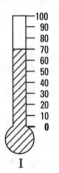
I

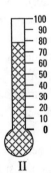
II

Figure I above shows the thermometer reading at 8 a.m. In Figure II, draw what the thermometer reading would have looked like at noon.

7. Consider each of the following. Does the equation show what the temperature was at 3 p.m. the same day?

yes	no	a)	Temperature $= 70 + 7(2)$
yes	no	b)	Temperature $= 70 + 5(2)$
yes	no	c)	Temperature $= 70 + 2 + 2 + 2 + 2 + 2 + 2 + 2$
yes	no	d)	Temperature $= 70 + 3(2)$
yes	no	e)	Temperature $= 70 + (8 + 3)2$

Analysis of Questions 4–7

An incorrect answer for question 4 may show that the student needs to read the problem more thoroughly. A student must be able to perform the basic level of qualitative analysis assessed by this question in order to be able to set up a workable solution plan for the problem. The teacher might want to discuss with the students the various quantities, known and unknown, in the problem.

Question 5 assesses whether a student can discriminate among the facts presented as well as bring some outside knowledge to the problem.

A representation of the information is introduced in question 6. This question, written in the free-response format, permits the teacher to assess whether the student is capable of correctly interpreting the relationship between the temperature at 8 a.m. and the temperature at noon. The student is solving, with a representation, a simpler form of the problem.

Question 7 presents two different ways to solve the problem correctly. Students who answer **yes** to 7b should be questioned to see whether they can determine the number of hours between 8 a.m. and 3 p.m. Those students who answer **yes** to 7d or 7e might not understand the role of the numbers 8 and 3 in the problem.

Guiding Class Discussion

The practice sheets and suggested activities in the Follow-up Instruction section under the concepts **Determine what is being asked, Identify or use assumed information relevant to a problem's solution, Use given information to set up a solution procedure, Represent given information by diagrams, charts, or tables**, and **Recognize possible alternate methods for a problem's solution** may be helpful for those students who are having difficulties with the ideas presented in these questions.

Focus of Questions 8–13

The group of questions 8–13 presents the idea of being paid time and a half, a situation that many students may encounter in their own lives. Question 8 focuses the student on what is being asked while questions 9–12 require that relevant given or assumed information be identified. In question 13 the given information is used to set up a solution procedure.

Questions 8–13

Questions 8–13 refer to the following story problem.

Carmen worked at a fast-food restaurant. For the first 8 hours she worked in a day, she was paid $4 per hour. For every hour after the first 8 hours, she was paid time and a half. One day Carmen worked 11 hours. How much was she paid? (Time and a half means one and one-half times the regular hourly wage.)

8. The answer will be a

   (A)   number of hours
   (B)   number of days
   (C)   dollar amount

9. "Carmen is paid time and a half."

Consider each of the following. In this problem, is the statement a correct interpretation of the sentence above?

⟨yes⟩ no **a)** Carmen is paid $6 an hour for every hour after the first 8 hours of work that day.

yes ⟨no⟩ **b)** Carmen works 8 hours plus one-half of 8 hours.

⟨yes⟩ no **c)** Carmen is paid $2 more an hour for every hour after the first 8 hours of work that day.

⟨yes⟩ no **d)** Carmen is paid $4 plus one-half of $4 an hour for every hour after the first 8 hours of work that day.

yes ⟨no⟩ **e)** Carmen is paid $4.50 an hour.

10. For how many hours was Carmen paid exactly $4 an hour?
    (A) 0
    (B) 3
    ⟨C⟩ 8
    (D) 11

11. Which of the following shows the total amount that Carmen earned for the first 8 hours?
    (A) $4
    ⟨B⟩ 8 × $4
    (C) 8 × $6
    (D) 8 × $4.50

12. For how many hours was Carmen paid time and a half?
    ⟨A⟩ 3
    (B) 4
    (C) 8
    (D) 11

13. Consider each of the following. Does the equation show how much Carmen earned that day?

⟨yes⟩ no **a)** 8(4) + 3(6) = Total pay in dollars

yes ⟨no⟩ **b)** 11(10) = Total pay in dollars

yes ⟨no⟩ **c)** 11(4) + 11(6) = Total pay in dollars

yes ⟨no⟩ **d)** 8(4 + 6) + 3(4 + 6) = Total pay in dollars

⟨yes⟩ no **e)** 11(4) + 3(2) = Total pay in dollars

Analysis of Questions 8–13

In question 8, the student should be able to relate the phrase "how much" in the last sentence of the problem to a dollar amount.

Question 9 presents several correct interpretations of time and a half. Understanding what time and a half means is necessary for solving the problem. While 9a, 9c, and 9d correctly apply the idea of time and a half, 9b and 9e do not. Students who answer **yes** to 9e may be interpreting the "half" in time and a half as half a dollar added to the regular hourly wage, or they may be dividing the sum of the amounts earned at the regular and overtime rates by 11 hours, which yields approximately $4.55. A **yes** answer to 9b may indicate that the student is confusing half of the salary rate with half of the hours. Question 9 points out the importance of being able to recognize when outside knowledge is needed to solve a problem. Your text or problems that the students write may provide further examples. The discussion could focus on various strategies for finding out how to use the information if the student does not have the outside knowledge needed. Some students may incorrectly answer the question because they simply do not know what time and a half means and do not understand the definition given.

Question 10 helps to assess whether a student can isolate the relevant information in the problem. It is important to realize that the rate of $4 an hour applies only to 8 of the 11 hours that Carmen worked. A student who answers **(B)** may be confusing the regular and overtime hours, while one who answers **(A)** or **(D)** either may not see the difference between the two rates of pay or may be ignoring the word "exactly."

Question 11 is an expansion of the ideas presented in questions 9 and 10. Once the student realizes that $4 an hour applies to only the first 8 hours, the dollar amount can then be determined for these hours. Teachers should check for consistency among the student's answers to questions 9, 10, and 11 to assess whether the student is able to follow the reasoning necessary for solving the problem.

Errors that the students make in question 12 may be similar to those they make in question 10. Teachers should check to see that the student's responses to questions 10 and 12 are consistent. Students who answer one question correctly and the other incorrectly may have only partial understanding of time and a half. Teachers may need to question the students further to see why they are providing inconsistent answers to these questions.

Question 13 asks the student to identify a correct method of solution. In 13b, the students need to see that working for 11 hours at $10 an hour is not the same as working for 8 hours at $4 an hour plus 3 hours at $6 an hour. Students answering **yes** to 13b may just be adding the numbers together without understanding how the numbers in the problem are related or the arithmetic involved in the calculation. A **yes** answer to 13c might mean that the student is unable to discriminate between the number of hours at each pay rate, while a **yes** answer in 13d might mean that the students do not understand that there is a difference in the pay rate for each number of hours. Part 13e may be difficult for some students to identify as correct since it involves a more sophisticated interpretation of time and a half. In this case, Carmen is paid the rate of $4 an hour for the entire 11 hours plus half of $4 (or $2) an hour for the 3 hours of overtime. It may be profitable to discuss with students other ways to represent Carmen's salary correctly in a numerical equation. For example, another correct equation is 8(4) + 3(4) + 3(2) = Total pay in dollars.

Guiding Class Discussion

For those students who are having difficulties or needing more practice with the ideas presented in these questions, the practice sheets and suggested activities under the concepts **Determine what is being asked, Identify given information relevant to a problem's solution, Use given information to set up a solution procedure**, and **Recognize possible alternate methods for a problem's solution** in the Follow-up Instruction section may be helpful. In particular, see **practice sheet 1** for exercises that focus on recognizing relevant and irrelevant information.

Focus of Questions 14–19

Questions 14–19 present another problem situation where the student must determine what is being asked, identify relevant given and assumed information, and use the information to set up a solution procedure. In addition, the reasonableness of the answer is also addressed.

While reading the situation presented with questions 14–19, one of the first things on which the student should focus is the word "average."

Questions 14–19

Questions 14–19 refer to the following situation.

Sue had several small bags of jelly beans. She counted the number of green jelly beans in each of them and obtained the following results: 5, 12, 5, 7, 9, 8. She wants to find the average number of green jelly beans in these bags.

14. What does Sue want to find?
- (A) The average number of jelly beans
- (B) The total number of jelly beans
- (C) The average number of green jelly beans
- (D) The total number of green jelly beans

15. How many bags of jelly beans did Sue have?
- (A) 5
- (B) 6
- (C) 7
- (D) 12

16. What is the *greatest* number of green jelly beans Sue found in any one of the bags?

Answer: *12* _____

17. What is the *least* number of green jelly beans Sue found in any one of the bags?

      Answer: *5*

18. The average number of green jelly beans in the bags will be between

    (A)   1 and 5
    (B)   5 and 12
    (C)   12 and 46

19. Consider each of the following. Does the statement *completely* describe how to find the average number of green jelly beans in the bags?

    yes (no)   **a)**   Add 5, 12, 5, 7, 9, and 8.
    yes (no)   **b)**   Add 5, 12, 5, 7, 9, and 8 and divide by 2.
    (yes) no   **c)**   Add 5, 12, 5, 7, 9, and 8 and divide by 6.
    yes (no)   **d)**   Add 5, 12, 5, 7, 9, and 8 and divide by 5.
    yes (no)   **e)**   Add 5 and 12 and divide by 2.

Analysis of Questions 14–19

A correct response to question 14 depends on a careful reading of the problem, since Sue wants to find the average number of *green* jelly beans and not just the average number of jelly beans in a bag.

Question 15 assesses whether students understand that they need to know the number of bags in order to find the average, and whether students see how this information can be obtained from the problem. The discussion of question 15 should focus on the idea that although relevant information is often not explicitly stated, it is available from the information that is given.

Questions 16 and 17 assess interpretation of the data presented in the problem. Note that the number 5 occurs twice in the data. Some students may misinterpret this to mean that 5 is the greatest number of green jelly beans (in question 16), since 5 is the only number that occurs twice. A discussion of exactly what the numbers in the problem represent would be helpful to these students.

Question 18 is a follow-up to 16 and 17. This question involves establishing boundary conditions on the answer. Before a student can estimate an average, the greatest and least numbers that are used in determining the average must be known. Estimating an answer is an important part of the preliminary analysis of a problem.

Question 19 presents several methods for finding the average number of green jelly beans in a bag. Part 19a is just the first step in a correct solution method for the problem. Students who answer **yes** to 19b and 19e, which both involve division by 2, may be misinterpreting average to mean to divide a sum in half. A **yes** response to 19d may indicate that the student does not realize that, since the number 5 occurs twice, it must therefore be counted twice in determining by what number to divide the sum. Students who incorrectly answer these questions need to understand the relationship between the average of a set of numbers and the range of values in that set of numbers. An average will never be less than the smallest number nor more than the greatest number. Using a number line to graph the average and the set of numbers given in the problem may be helpful.

Guiding Class Discussion

The practice sheets and suggested activities in the Follow-up Instruction section under the concepts **Determine what is being asked, Identify given information relevant to a problem's solution, Use given information to set up a solution procedure**, and **Estimate and determine the reasonableness of a problem's solution** may be helpful for those students who are having difficulties or needing more practice with the ideas presented in these questions.

Focus of Question 20

Question 20 presents the idea that there may be alternate solution methods for a problem.

Question 20

20.

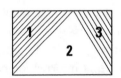

Consider each of the following. Does the statement completely describe how to find the area of the shaded region in the figure above?

(yes)	no	**a)**	Add the areas of triangle 1 and triangle 3.
yes	(no)	**b)**	Add the areas of all three triangles.
(yes)	no	**c)**	Find the area of the rectangle and then subtract the area of triangle 2.

Analysis of Question 20

Students may answer 20a correctly but not answer 20c correctly, or vice versa. Although it is not necessary that students always recognize multiple ways to solve a problem, this question provides a good example of a simple problem that students should be able to solve easily using different methods.

Guiding Class Discussion

For those students who are having difficulties or needing more practice with the ideas presented in this question, the practice sheets and suggested activities in the Follow-up Instruction section under the concepts **Use given information to set up a solution procedure** and **Recognize possible alternate methods for a problem's solution** may be helpful.

Focus of Questions 21–23   The free-response format of questions 21–23 helps the teacher to assess the student's ability to interpret information given in a diagram, as well as the student's ability to formulate different ways to solve a problem. Students who are unable to begin to answer these questions may need help initiating their problem-solving strategy. The teacher should question the students to make certain that they understand what is being asked.

Questions 21–23

21.

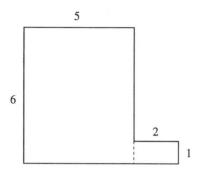

Show by drawing lines on the figure above how the figure can be divided into two rectangles.

22.

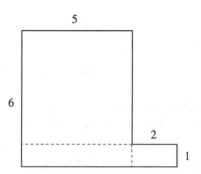

Show by drawing lines on the figure above how the figure can be divided into two rectangles and a square.

23. Area of a rectangle = length × width

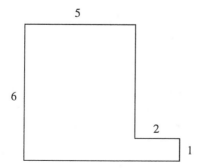

One way to write an equation that gives you the area of the figure above if it is divided into two rectangles is

$$(5 \times 6) + (2 \times 1) = 32.$$

Write another equation that also gives you the area of the figure if it is divided into two rectangles and a square.

Answer: $(5 \times 5) + (5 \times 1) + (2 \times 1) = 32$

Analysis of Questions 21–23

Questions 21 and 22 ask the student to show the exact position of the lines drawn. If students do not suggest the given solutions, the teacher may want to encourage the students to find alternate ways of representing the given figure as sums or differences of rectangles. Students do not have to be able to see every possible way to solve a problem. The purpose of these exercises is to help them see that if one solution method does not work, others can be found. Another method of dividing the figure is given below.

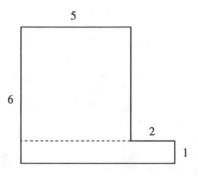

Guiding Class Discussion

To answer question 23, the student may draw additional lines on the diagram. Encourage the students to explain in their own words the given diagram and the accompanying equation $(5 \times 6) + (2 \times 1) = 32$. Questions such as the following may be helpful.

Where did 5×6 come from?
Where did 2×1 come from?
Why are these two products added together?
What does 32 represent?
How was the diagram divided to form the equation?

It should be pointed out to students that a seemingly complicated problem such as this is manageable when broken into smaller steps.

For those students who are having difficulties or needing more practice with the ideas presented in these qustions, the practice sheets and suggested activities in the Follow-up Instruction section under the concept **Represent given information by diagrams, charts, and tables** may be helpful. In particular, **practice sheet 2** provides more figures that can be used for practice in dividing areas up into rectangles.

Focus of Questions 24–25

Students' responses to questions 24 and 25 will demonstrate whether they can carefully read, interpret, and recognize what information is relevant and how it is to be used to solve the problem. The students' diagrams should show a square with side of length 10 and a border inside the square of width 1. The diagrams should also be labeled with the appropriate numbers.

Questions 24–25

Questions 24–25 refer to the following situation.

Chris wants to install a 1-foot-wide tile border around the edge of the floor of a square room that measures 10 feet on each side. At least one side of each tile is to be placed against a wall of the room.

24. Draw a diagram that represents what the floor might look like after Chris has installed the tile border. Label the diagram with appropriate measurements of the floor and tile.

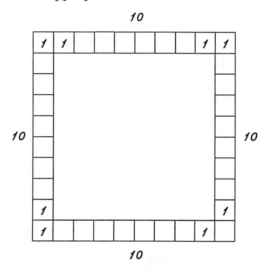

25. Describe with words and/or numbers and symbols how to find the total area Chris is going to cover with the tile.

Answer: *Subtract the area of the square with side of length 8 from the area of the square with side of length 10, (10 × 10) − (8 × 8)*

Analysis of Questions 24–25

Question 25 asks the student to describe how to find the area that will be covered by the tile. The diagram that the student draws in response to question 24 is meant to aid the student in answering this question. It is possible to arrive at the solution in several ways. One way is to label the figure as follows.

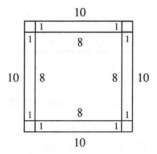

Area to be covered with tile $= (10 \times 10) - (8 \times 8) = 36$.

Another solution method could be to use the equation

Area to be covered with tile $= (4 \times (1 \times 1)) + (4 \times (8 \times 1)) = 36$.

Some students may write the description as an English sentence rather than an equation.

A third solution method could be to draw all the squares on the border of the 10×10 square (as shown below) and count them to arrive at the answer 36.

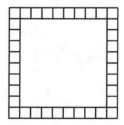

Note that the reasoning necessary to solve this problem is similar to that needed to work with questions 21, 22, and 23.

Guiding Class Discussion

The practice sheets and suggested activities in the Follow-up Instruction section under the concepts **Use given information to set up a solution procedure** and **Represent given information by diagrams, charts, or tables** may be helpful for those students who are having difficulties or who need more practice with the ideas presented in these questions. **Practice sheet 3**, in particular, can be used to develop skill in drawing and describing models that could be used to represent mathematical situations.

Focus of Questions 26–27

Questions 26 and 27 involve distance, rate, and time. Question 26 assesses whether the student can understand what is being asked. The student should be able to understand that the phrase "how fast" indicates a rate. Question 27 assesses whether the student can estimate the value of the answer. Students need to understand that complex calculations are *not* necessary to establish boundary conditions on the answer.

Questions 26–27

26. It takes 20 seconds ($\frac{1}{3}$ minute) for a person to travel from the bottom to the top of a certain escalator. The escalator is 30 feet long. How fast is the escalator moving?

The answer to this problem will be a

(A) distance
(B) time
(C) rate ⟵ circled

27. Another escalator moves at 200 feet per minute. If it takes $\frac{1}{5}$ minute for a person to travel from the bottom to the top of the escalator, how long, in feet, is the escalator?

The answer to this problem will be between

(A) 200 feet and 1,000 feet
(B) 20 feet and 200 feet ⟵ circled
(C) 5 feet and 20 feet

Guiding Class Discussion

The teacher may want to review the distance-rate-time formula with the students. Students may find it helpful to construct their own rate problems. Encourage them to write number sentences or use tables to show the relationships between quantities in the problems they construct.

Question 27 also encourages the development of estimation techniques. The teacher might show the students how to use relative sizes of fractions to perform such estimations; for example the solution to question 27 would be greater than $\frac{1}{10} \times 200$ or 20 feet, since $\frac{1}{5} > \frac{1}{10}$ and the solution should be less than $\frac{1}{2} \times 200$ or 100 feet, since $\frac{1}{5} < \frac{1}{2}$.

For those students having difficulties or needing more practice with the ideas presented in these questions, the practice sheets and suggested activities in the Follow-up Instruction section under the concepts **Determine what is being asked** and **Estimate and determine the reasonableness of a problem's solution** may be helpful.

Focus of Questions 28–29

Questions 28 and 29 present a setting frequently used in pre-algebra and algebra word problems. Students should be encouraged to use a table to show relationships between Joe's age and Ann's age at various points in time. Tables can be a useful way for students to organize information given in a problem.

Question 29 asks the student to write another story for the situation presented in question 28 in order to assess further the student's understanding of the information given.

Questions 28–29

28. Joe is 10 years old and Ann is 13 years old. In 4 years Joe will be 14. How old will Ann be then?

Fill in the table below using the information given in the problem above.

Joe	Ann	
8	*11*	Age 2 years ago
10	13	Age now
14	*17*	Age in 4 years

29. Use the information in question 28 to write another story problem about Joe and Ann.

Answer: *Joe is 3 years younger than Ann. If Joe is 10 years old, how old was Ann 2 years ago?*

Guiding Class Discussion

The teacher may want to give a sample question similar to question 29 in class so that the students understand how to answer the question. The teacher should focus the students' attention on finding the relationship between the quantities, using this relationship to generate different situations.

The following are some other sample responses for question 29.

1. Ann is 3 years older than Joe. In 4 years Joe will be 14 years old. How old will Ann be then?

2. Joe is 3 years younger than Ann. If the sum of their ages is 23, how old was Joe two years ago?

The practice sheets and suggested activities in the Follow-up Instruction section under the concepts **Represent given information by diagrams, charts, or tables** and **Recognize possible alternate methods for a problem's solution** may be helpful. Specifically, **practice sheets 4** and **5** focus on the use of tables to organize information and solve problems while **practice sheets 5, 6,** and **7** present problems that are intended to broaden the student's view of problems and problem solving. Practice sheets 4–7 give opportunities to practice the skills developed in this unit and to develop a willingness to try new approaches to problem solving. In particular, practice sheets 5 and 6 focus on problems that involve logical reasoning rather than numerical calculation. Practice sheet 7 asks students to conduct a survey on a topic of interest to them.

Follow-up Instruction

After the Instructional Assessment has been used to pinpoint students' conceptual weaknesses, the teacher may wish to use the following suggested activities and practice sheets to help correct those misconceptions. Since every word problem necessarily involves a number of concepts, activities in this unit often involve many concepts; practice sheet 7, for example, summarizes all of the concepts in the unit.

Determine what is being asked

This concept is specifically addressed by **practice sheet 7** which involves the students in conducting a survey. When the students choose a topic for their survey, they are defining the problem for the survey, thus determining what is being asked and what form the answer should take. (Practice sheet 7 is discussed more fully under the label All unit concepts.)

Activity. To give students more practice in determining what is being asked in a problem, the teacher may select word problems from the regular text or other sources and ask students to identify what form the answer should take (*i.e.*, a rate, time of day, amount of money, number of hours or other quantity, distance, an age, an average, etc.). A more challenging activity is to tell the students what form the answer is to take and have them write a corresponding problem. Having the students work in pairs might make this task easier for them.

In the Follow-up Assessment, questions 4, 8, 14, and 26 address this concept.

Identify given information relevant to a problem's solution

Practice sheet 1 consists of several simple problems that include both relevant and irrelevant information to give students practice in weeding out the irrelevant information.

Practice sheet 5 helps develop the student's ability to identify given information relevant to a problem's solution and to identify or use assumed information relevant to the problem's solution. This activity also illustrates how a table can serve as an effective way to organize information and analyze the given situation.

Encourage the students to use a chart in their solution to organize the information presented. The solution of the problem on practice sheet 5 can be facilitated by the use of a chart like the one below.

	Alice	Brent	Candy	Dale
Track				
Gymnastics				
Swimming				
Basketball				

Students can mark the table to show what each piece of information tells them.

Since Candy is the sister of the gymnastics participant, Candy's sport is not gymnastics. Alice's sport must be swimming, since her sport is the only one that requires water. A 400-meter race might be either a track or a swimming event, but since Alice's sport is swimming, Dale must run track. Therefore by elimination, Brent's sport is gymnastics and Candy's sport is basketball.

For the second problem on practice sheet 5, drawing a diagram or table to represent the seven bins will help students organize their solutions. They may begin by filling in the bins that always contain a particular vegetable. For example, bin 1 always has lettuce. Bin 2 always has tomatoes since tomatoes must be next to lettuce. Bin 7 always has carrots. The problem becomes how to fill bins 3–6 which can be done by remembering the other restrictions that onions are never next to potatoes and broccoli is never next to green beans. Students may suggest various solution methods. Encourage discussion of the alternate methods. This problem could also be used as a group activity where each group finds as many arrangements as possible.

One possible arrangement is lettuce, tomatoes, broccoli, potatoes, beans, onions, and carrots.

One source of student errors in solving word problems is the incorporation of incorrect assumptions. Focusing the discussion of practice sheet 5 on the role of assumptions in processing information by asking students to explain any assumptions they made may help students realize their errors.

Practice sheet 6 helps develop the student's ability to identify given information relevant to a problem's solution and to recognize possible alternate methods for a problem's solution.

This activity presents an opportunity for the students to work in pairs. The teacher should encourage students to consider all plausible ways of solving the problem. Also, the students can practice identifying relevant and irrelevant information given in the problem. They should also identify any assumptions they have made.

The following classroom discussion and activity are intended to get students to realize that problem-solving requires planning and that a part of the planning is processing the necessary information. Situations with which the students are familiar will be more meaningful to them and should aid in their understanding of the various aspects of a problem.

Activity. Ask the students what types of situations require advance planning and what kind of information needs to be known in order to develop such plans. For example, a task such as choosing what type of clothing to wear on a certain day requires knowing what one will be doing (working, playing a sport, going to school, etc.) and what the weather conditions will be. These latter points are examples of information that is relevant to solving a problem.

Another situation is that of planning to save for a large purchase, such as a new bicycle, a moped, or a stereo system. Considerations in this instance could include how much the item costs, how much can be saved per week to meet the goal, and for how many weeks one needs to save this amount.

Activity. After discussing these examples or others provided by the students, students could be asked to use such resources as books, magazines, newspapers, or television to find examples of events or projects of interest to them which required advance planning. Events may be historical, current, or personal and include, for example, famous battles, scientific explorations or experiments, tournaments for sports, music or other festivals, family vacations, or social events at school. Students should focus on the kinds of information needed to make the plans. A bulletin board display prepared by the students could group events by type of event or type of information needed to develop the plans.

In the Follow-up Assessment questions 3, 10, 16, and 17 address this concept.

Identify or use assumed information relevant to a problem's solution

In solving the reasoning problem on **practice sheet 6**, students are likely to make and use various assumptions. Since the assumptions made can greatly influence the solution, students should be encouraged to state their assumptions and explain how their assumptions influence their solution. Practice sheet 6 is discussed more fully under the concept Identify given information relevant to a problem's solution.

In conducting their survey for **practice sheet 7**, students may also make certain assumptions concerning the people they interview, the people's understanding of the question or questions asked, or other conditions. In the early planning stages of the survey, assumptions should be explicitly addressed. The data analysis and survey report may reveal that the students also made other assumptions not previously identified. Practice sheet 7 is discussed more fully under the label All unit concepts.

In the Follow-up Assessment questions 5 and 9 address the concept.

Use given information to set up a solution procedure

Practice sheet 6 presents a situation in which a great deal of information is given. Since the descriptive nature and level of detail in the problem may be unfamiliar to students, the teacher may want to encourage discussion about how approaches to this type of problem are the same as and different from approaches to numerical problems. (Practice sheet 6 is discussed more fully under the concept Identify given information relevant to a problem's solution.)

Practice sheet 7 addresses the concept of using given information to set up a solution procedure when students collect and organize their survey data. For more discussion of practice sheet 7 see the All unit concepts section.

In the Follow-up Assessment the related questions are 2, 7, 11, 12, 13, 15, 19, 20, 23, and 25.

Estimate and determine the reasonableness of a problem's solution

Activity. An activity that helps to develop the ability to estimate and check the reasonableness of a problem's solution is to have two students simulate a buyer-seller relationship in which one student assumes the role of buyer. The teacher could list several items by name and their corresponding unit prices on the chalkboard. Have the students determine how much money is needed to buy certain items to give them practice in establishing upper boundaries on answers.

In the Follow-up Assessment questions 1, 18, and 27 address this concept.

Represent given information by diagrams, charts, or tables

As a follow-up to questions 21 and 22 in the Instructional Assessment, **practice sheet 2** gives additional figures. Students who have difficulty determining lengths of unlabeled line segments in the diagram could be given more practice with such exercises. By drawing additional lines, students may be better able to see how to determine unlabeled measures. These types of problems are a good way to show students the existence of different types of solution methods for one problem. If desired, the teacher can also give the students dimensions for the figures in practice sheet 2 and ask them to find each area. Note that students may wish to complete the figures and approach the problem by subtraction. This gives the teacher another opportunity to point out that a problem may have more than one solution.

The focus of **practice sheet 3** is on effective communication. In describing a design, the students concentrate on the design's components and relationships among those components. Thus, they begin to develop the ability to understand and design models to represent mathematical situations.

The object of the activity is for each student to give accurate directions to his or her partner that will enable the partner to draw the figure. The student who is drawing may not ask questions and the describer may not point or gesticulate. After the student who is drawing completes the figure using the describer's directions, the students can check their creations with the original drawing. The students can then switch roles and repeat the process using a second figure. Have the student who is drawing the picture tell his or her partner what was confusing or helpful about the description. All the partners working with one drawing could work in a group to create the best way to describe the figure. A representative from the group (or the teacher) could read the description to the other half of the class to see if they can accurately draw the figure and show that the description is helpful. In this activity simple geometric figures such as circles, ovals, triangles, rectangles, and squares are used so that the students are able to describe easily not only the shape of each geometric figure, but also how the figures are oriented with respect to each other. After the students have developed some confidence in this activity, it might be possible to incorporate more complex figures into the exercise.

By using tables to show how the numbers in a problem are related to each other, **practice sheet 4** helps students understand the reasoning used to construct a word problem. The comments below show how the technique of using tables also allows opportunities to extend assessment and instruction.

In problem 1 on practice sheet 4, the sum could be 156, 159, or 162. This problem presents many opportunities for discussion. The teacher could pose such questions as: Why are there three possible answers? How many possible answers would there be if two of the three consecutive whole numbers were given? If two of the three numbers *and* the sum of the three numbers are given, how many possibilities are there for the third (unknown) number?

In problem 2, the least amount is $13. The problem could then be modified to further utilize the table. For example, the students could be asked "Is it possible to have a total of $28 in $1 and $5 bills in the drawer?" and/or "If the number of $1 bills is twice the number of $5 bills, what is the total value of the $1 and $5 bills in the drawer?"

For problem 3, the other addend could be 38, 26, 14, or 2. This problem presents an opportunity to discuss divisibility and factors. It also highlights subtraction as finding the missing addend.

As a follow-up to practice sheet 4, a helpful class exercise might be to have groups of students construct several story problems and then exchange and solve them. Encourage students to use a variety of settings for their problems (number, rates, mixtures, money, and geometry, to name a few).

In the Follow-up Assessment questions 6, 21, 22, 24, and 28 address this concept.

Recognize possible alternate methods for a problem's solution

Practice sheet 2 (discussed under the concept Represent given information by diagrams, charts, or tables), **practice sheet 6** (discussed under the concept Identify given information relevant to a problem's solution), and **practice sheet 7** discussed below all relate to this concept.

In the Follow-up Assessment questions 7, 13, 20, 23, and 29 are pertinent.

All unit concepts

Practice sheet 7 ties together all the concepts and abilities that are addressed in this unit. It illustrates the importance of including these concepts and abilities as components of good problem-solving strategies.

The activity is based on a survey.[1] The objective of the survey is determined by the class or teacher and should be something that is of interest to the students in the class. For example, a poll could be conducted to determine the favorite musical group, favorite type of food, number of hours of recreational time per week, or amount of weekly allowance. The advantage of using a survey as an activity is that it is extremely flexible and can be tailored to the specific interests of the class and the needs of the teacher.

After making the survey, the students should be encouraged to make up story problems using data from the survey.

The chart below shows the relationship between the different components of implementing a survey and the concepts and abilities needed to solve word problems.

Steps in Survey	Concepts and Abilities
Choose a topic	Determine what is being asked
Plan what information is needed and how to get it	Processing information
Collecting data	Different ways to solve a problem
Organize data	Processing information Representing relevant information
Interpret data	Processing information
Use data to estimate and predict	Establish boundary conditions on answers

[1] *The Arithmetic Teacher* 32.9 (May 1985): 10–12.

Bibliography of Resource Materials

Charles, R., et al. *Problem-Solving Experiences in Mathematics.* Menlo Park, CA: Addison-Wesley, 1985.

Crandall, Jody, et al. *English Skills for Algebra.* Center for Applied Linguistics, Washington, D.C., Englewood Cliffs, NJ: Prentice-Hall, 1987.

Giblin, Peter, ed. *Mathematical Challenges: Puzzles and Problems in Secondary Mathematics.* Providence, RI: Janson Publications, Inc., 1989.

Greenes, Carole, et al. *Mathletics: Gold Medal Problems.* Providence, RI: Janson Publications, Inc., 1990.

Hanson, Merrily P., and Maryann Marrapodi. *Practicing Problem-Solving.* New York: Garber Group, Random House, 1982.

Harnadek, Anita. *Math Mind Benders.* Pacific Grove, CA: Midwest Publications.

———. *Mind Benders B-level Books.* Pacific Grove, CA: Midwest Publications.

———. *Mind Benders Software.* Computer software. Brooke Boering, programmer. Midwest Publications. Apple II+/IIe/IIc/IIGS, 64K.

Krulik, Stephen, and Jesse A. Rudnick. *Problem Solving: A Handbook for Teachers.* Newton, MA: Allyn and Bacon, 1980.

———. *A Sourcebook for Teaching Problem Solving.* Newton, MA: Allyn and Bacon, 1984.

Mason, Burton, and Stacey Mason. *Thinking Mathematically.* Menlo Park, CA: Addison-Wesley, 1985.

Sloyer, Cliff. *Fantastiks of Mathematiks: Applications of Secondary Mathematics.* Providence, RI: Janson Publications, Inc., 1986.

Answers

Answers to Instructional Assessment

1. a) yes
  b) no
  c) yes
  d) yes
  e) no
  f) yes
2. a) no
  b) yes
  c) yes
  d) no
3. a) no
  b) no
  c) yes
  d) no
  e) yes
  f) no
  g) yes
4. A
5. C
6.

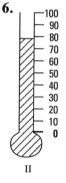

7. a) yes
  b) no
  c) yes
  d) no
  e) no
8. C
9. a) yes
  b) no
  c) yes
  d) yes
  e) no
10. C
11. B
12. A
13. a) yes
  b) no
  c) no
  d) no
  e) yes
14. C
15. B
16. 12
17. 5
18. B
19. a) no
  b) no
  c) yes
  d) no
  e) no

20. a) yes
  b) no
  c) yes
21.

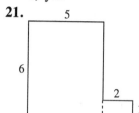

22.

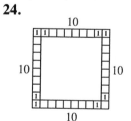

23. $(5 \times 5) + (5 \times 1) + (2 \times 1) = 32$
24.

10
10   10
10

25. Subtract the area of the square with side of length 8 from the area of the square with side of length 10; $(10 \times 10) - (8 \times 8)$
26. C
27. B
28. 8, 11; 14, 17
29. Answers will vary. For example, Joe is 3 years younger than Ann. If Joe is 10 years old, how old was Ann 2 years ago?

179

Answers to Practice sheets

Practice sheet 1

1. Irrelevant information: "Three weeks ago she bought a book for $7.95."
2. Irrelevant information: "The capacity of the gas tank is 1.5 gallons."
3. Irrelevant information: "there are 9 boys and 6 girls."
4. Irrelevant information: "Earrings were on sale for $8.95 a pair."
5. Answers will vary.

Practice sheet 2

Answers will vary. For example,

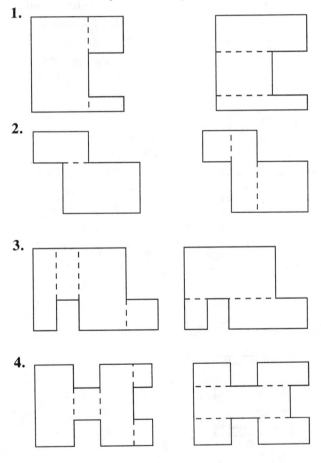

1.
2.
3.
4.

Practice sheet 3

1. Draw a square. Draw a second square of the same size that has its upper left corner at the center of the first square and has its sides parallel to the first square. Shade the area that both squares share.

2. Draw a square. Inside the square draw a circle that touches all four sides of the square. Inside the circle draw a square so that its four corners touch the circle and its sides are parallel to the first square. Inside the smaller square draw a circle that touches the four sides of the smaller square.

3. For example: Draw a rectangle so that the longer sides are horizontal. Extend the left side above the rectangle and the right side below the rectangle. Connect the ends of the extensions to the end of the closest long side.

4. For example: Draw a circle. Inside the circle draw a triangle with equal sides whose vertices touch the circle. Position the triangle so that one vertex is pointing north. Draw a line from the left-most vertex to the middle of the opposite side of the triangle.

5. For example: Draw a rectangle so that its longer sides are horizontal. Draw the diagonal from the lower left corner to the upper right corner. Form another rectangle by drawing a line straight down from the center of the top side of the rectangle down to the diagonal and another line straight across from the center of the right side of the rectangle to the diagonal.

6. For example: Draw a square. Draw a horizontal line through the center of the square. Extend the line an equal distance on each side. Connect the ends of the extensions to the closest corners of the square.

7. Draw a square. Inside the square draw a circle that touches all four sides of the square. Shade in the upper right quarter of the circle. Using the center of the right side of the square as the upper left corner, draw a second square of the same size as the first square. Shade in the lower right fourth of the second square.

8. For example: Draw a rectangle whose longer sides are vertical. Extend the upper side to the right of the rectangle and the lower side to the left of the rectangle the same distance. Connect the endpoints of the extensions to the most distant corner (vertex) of the rectangle.

Practice sheet 4

1. Three Consecutive Whole Numbers	Sum
53, 54, 55	162
52, 53, 54	159
51, 52, 53	156

2. Numbers Less Than 50 That Are Divisible by 3 and 4	Number Which When Added Gives a Sum of 50
12	38
24	26
36	14
48	2

3. Number of $1 Bills	Number of $5 Bills	Total Value
1	8	$41
2	7	$37
3	6	$33
4	5	$29
5	4	$25
6	3	$21
7	2	$17
8	1	$13 least amount

Practice sheet 5

1. Candy–basketball
Alice–swimming
Dale–track
Brent–gymnastics

2. Possible arrangements:

lettuce, tomatoes, broccoli, potatoes, beans, onions, carrots

lettuce, tomatoes, broccoli, onions, beans, potatoes, carrots

lettuce, tomatoes, beans, potatoes, broccoli, onions, carrots

lettuce, tomatoes, beans, onions, broccoli, potatoes, carrots

lettuce, tomatoes, potatoes, broccoli, onions, beans, carrots

lettuce, tomatoes, potatoes, beans, onions, broccoli, carrots

lettuce, tomatoes, onions, beans, potatoes, broccoli, carrots

lettuce, tomatoes, onions, broccoli, potatoes, beans, carrots

Practice sheet 6

One possible solution to the problem is that the hiker plugged up the hole in the floor of the pit with the rock, allowed the water level to rise, and used the boards as a raft to float out of the hole.

Another solution might be that the hiker could have climbed out if the tree was tall and sturdy enough and the branches extended close enough to the top edge of the pit.

Practice sheet 7

Answers will vary.

Answers to Follow-up Assessment

1. a) yes
   b) yes
   c) no
   d) no
   e) no
2. a) no
   b) no
   c) yes
   d) yes
3. a) yes
   b) yes
   c) no
   d) no
   e) yes
   f) no
4. D
5. C
6.

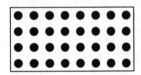

7. a) yes
   b) no
   c) no
   d) no
   e) no
8. A
9. a) yes
   b) yes
   c) yes
   d) no
   e) no

10. C
11. B
12. B
13. a) yes
   b) no
   c) no
   d) yes
   e) no
14. B
15. B
16. 26
17. 19
18. B
19. a) no
   b) no
   c) no
   d) yes
   e) no
20. a) yes
   b) no
   c) yes
   d) yes
21.

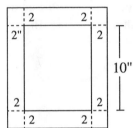

22.

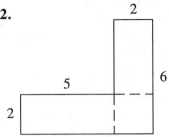

23. $(2 \times 5) + (2 \times 4) + (2 \times 2)$
    $= 22$

24.

25. Subtract the area of the picture from the area of the picture and frame.
26. C
27. A
28.

Steve	Denise	
11	8	Age 3 yrs. ago
14	11	Age now
19	16	Age in 5 yrs.

29. For example: Steve is 3 years older than Denise. If Denise is 11 years old, how old will Steve be in 5 years?

ALGE**BRIDGE**™

Concept

of Variable

Concept of Variable

Introduction

A key to understanding algebra and a major obstacle for many students is understanding the concept of variable. Students who study algebra have trouble understanding what letters such as x and y mean in expressions, equations, and inequalities, and why variables sometimes do not vary. Students' confusion may be the result of the generalization of the term by teachers and others who use mathematics. In algebra, a variable may refer to a symbolic constant (*e.g.*, the k in the equation $x = k$ where k is a constant), a parameter (*e.g.*, the t in the system of equations $x = t + 3$ and $y = 4t$), an unknown (*e.g.*, the x in $4x + 20 = -12$), or an unconstrained variable (*e.g.*, the x or y in $8x + 3y = 15$).

In this unit, different roles of a variable will be presented. They are

– the use of a variable as an unknown in an expression;

– the use of a variable as a specific unknown in a linear equation;

– the use of a variable in a linear inequality;

– the use of a variable as a pattern generalizer;

– the use of a variable in formulas and word problems.

Since the purpose of this unit is to help students to develop an understanding of the concept of variable and not to develop the ability to solve an equation or an inequality, solutions to problems presented in the unit are obtained using substitution. This use of substitution also reinforces the idea that a variable can represent a number or a set of numbers, depending on the context of the problem.

Cross-References to
Pre-Algebra and Algebra
Textbook Topics

The concepts and abilities addressed in this unit are not often treated explicitly in pre-algebra or algebra textbooks. Variables appear throughout the textbooks, however, and the use of this unit would be helpful when students are introduced to them.

This unit may also be helpful when studying the following topics that appear in many pre-algebra and algebra textbooks.

algebraic expressions	**linear equations**	**variables**
formulas	**linear inequalities**	**variation**
functions and graphs	**ratio and proportion**	**word problems**

List of Terms

The following is a list of words and phrases that are frequently used in this unit. The teacher may want to add other words and phrases to this list.

area	**equation**	**is a name for**	**replaced by**
corresponding	**expression**	**is greater than**	**rule**
depends on	**formula**	**is less than**	**statement**
digit	**inequality**	**perimeter**	**table**
			variable

Classification by Concept

Target Concepts and Abilities	Questions in Follow-up Assessment	Questions in Follow-up Assessment	Practice Sheets
Use of a variable as an unknown in an expression	1, 2, 3, 4, 8, 9, 10, 11, 12, 13, 14	1, 2, 3, 4, 8, 9, 10, 11, 12, 13, 14	1, 2, 4
Use of a variable as a specific unknown in a linear equation	5, 6, 7, 15	5, 6, 7, 15	3
Use of a variable in a linear inequality	16, 17, 18, 19	16, 17, 18, 19	
Use of a variable as a pattern generalizer	20, 21, 22, 23	20, 21, 22, 23	5
Use of a variable in formulas and word problems	24, 25, 26, 27, 28	24, 25, 26, 27, 28	6

Classification by Question

Target Concepts and Abilities	Use a variable as an unknown in an expression	Use a variable as a specific unknown in a linear equation	Use a variable in a linear inequality	Use a variable as a pattern generalizer	Use a variable in formulas and word problems
Question Number or Practice Sheet Number 1	IA FA PS				
2	IA FA PS				
3	IA FA	PS			
4	IA FA PS				
5		IA FA	PS		
6		IA FA		PS	
7		IA FA			PS
8	IA FA				PS
9	IA FA				PS
10	IA FA				
11	IA FA				
12	IA FA				
13	IA FA				
14	IA FA				
15		IA FA			
16			IA FA		
17			IA FA		
18			IA FA		
19			IA FA		
20				IA FA	

IA - Questions in Instructional Assessment
FA - Questions in Follow-up Assessment
PS - Practice Sheets

Classification by Question

Target Concepts and Abilities	Use a variable as an unknown in an expression	Use a variable as a specific unknown in a linear equation	Use a variable in a linear inequality	Use a variable as a pattern generalizer	Use a variable in formulas and word problems
Question Number or Practice Sheet Number 21	_____	_____	_____	IA FA	_____
22	_____	_____	_____	IA FA	
23	_____	_____	_____	IA FA	
24	_____	_____	_____	_____	IA FA
25	_____	_____	_____	_____	IA FA
26	_____	_____	_____	_____	IA FA
27	_____	_____	_____	_____	IA FA
28	_____	_____	_____	_____	IA FA

IA - Questions in Instructional Assessment
FA - Questions in Follow-up Assessment
PS - Practice Sheets

Instructional Assessment

In the Instructional Assessment, the concept of variable is introduced first in a numerical context by having the student identify different numerical expressions that are each equal to a particular number. Letters are used to represent variables after the student has been introduced to the use of the □ as a placeholder. Since variables are encountered in algebraic expressions, equations, and inequalities, this unit explores their use in each of these settings. Students are also introduced to the use of variables as pattern generalizers and in formulas and word problems.

In this unit, only whole numbers are used for the replacement values of the variables. This will permit the teacher to concentrate on the development of the concept itself. When the students have gained an understanding of the concept of variable, the teacher may want to use fractions and integers as replacement values for variables. Teachers may find it useful to review with students the concepts presented in this unit at appropriate points in the curriculum to reinforce understanding of these concepts.

If there is not sufficient time to review examples of the different question types, review the directions below with the students before they work on the *Algebridge*™ exercises.

Directions

There are two types of questions in this exercise.

The first type gives you several choices related to a specific question. You must answer **yes** or **no** for each of these choices.

The second type asks you to write your own answer to the problem.

Focus of Question 1

The purpose of question 1 is to assess the student's understanding of the idea that a number can be expressed in many different ways. This is the central idea that underlies the concept of variable.

---

1. Consider each of the following expressions. Is it a name for 8?

(yes) no	**a)**	$5 + 3$
(yes) no	**b)**	$11 - 3$
(yes) no	**c)**	$\dfrac{16}{2}$
(yes) no	**d)**	$4 \cdot 2$
yes (no)	**e)**	$\dfrac{2}{16}$
yes (no)	**f)**	$3 - 11$
(yes) no	**g)**	$3 + 5$
(yes) no	**h)**	$2 \cdot 4$

Analysis of Question 1

The student must first realize that $5 + 3$ and $1 + 7$ are both names for 8 in order to understand later that $\square + 3$ is a name for 8 when 5 is placed in the $\square$ and $1 + y$ is a name for 8 when 7 replaces y.

A related concern in this question is whether students have a correct understanding of the commutativity of operations. Students frequently ignore the order in which subtraction and division are written in a problem. A student who answers **yes** to questions 1b and 1f or **yes** to questions 1c and 1e may not understand that the operations of subtraction and division are not commutative. This response pattern may also help explain student errors in questions 2 and 5.

Guiding Class Discussion

The practice sheets and suggested activities in the Follow-up Instruction section under the concept **Use of a variable as an unknown in an expression** may be helpful for those students having difficulty or needing more practice. In particular, **practice sheets 1 and 2** help students to build a foundation for the concept of variable by asking them to express given numbers in several ways. Practice sheet 2 may be more challenging since it places restrictions on the digits that may be used in the expression.

Focus of Questions 2–4

Questions 2, 3, and 4 develop the use of the $\square$ as a placeholder in an expression. The teacher should discuss the use of substitution as a way to determine the number that should be placed in the $\square$.

Questions 2–4

2. Consider each of the following expressions. Is it a name for 8?

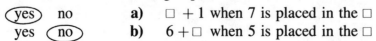

(yes) no	**a)**	$\square + 1$ when 7 is placed in the $\square$
yes (no)	**b)**	$6 + \square$ when 5 is placed in the $\square$

(yes) no   **c)** $\square - 10$ when 18 is placed in the $\square$

(yes) no   **d)** $1 + \square$ when 7 is placed in the $\square$

yes (no)   **e)** $10 - \square$ when 18 is placed in the $\square$

3. Consider each of the following numbers. If the number is placed in the $\square$, will $\square + 15$ be a name for 37?

     yes (no)   **a)**   12

     yes (no)   **b)**   15

     (yes) no   **c)**   22

     yes (no)   **d)**   37

     yes (no)   **e)**   52

4. If each expression below is a name for 8, what number should be placed in the $\square$?

     a) $\square + 3$;   _5_   should be placed in the $\square$.

     b) $\square - 3$;   _11_   should be placed in the $\square$.

     c) $\dfrac{\square}{8}$;   _64_   should be placed in the $\square$.

     d) $26 - \square$;   _18_   should be placed in the $\square$.

     e) $\dfrac{16}{\square}$;   _2_   should be placed in the $\square$.

Analysis of Questions 2–4

In question 2 the student must realize that 8 can be named by more than one expression involving the $\square$ as long as the appropriate number is placed in the $\square$. Students who answer **yes** to 2a and **no** to 2d, or vice versa, may not understand that the order in which two numbers are added does *not* affect the result, while students who answer **yes** to 2e may not understand that the order in which two numbers are subtracted *does* affect the result. It is important that students realize which arithmetic operations are commutative so that they are able to understand operations in the context of solving equations.

While the values for the $\square$ are given for each expression in question 2, question 3 suggests several possible numbers to place in the $\square$ when the given expression is a name for 37. If the student answers **yes** to 3a, 3b, 3d, or 3e in addition to 3c, he or she must be made aware that there is only *one* number that can be placed in the $\square$ for which $\square + 15$ will name 37.

Question 4 presents the student with the opportunity to determine the number that should be placed in the $\square$ when told that the given expression is a name for 8. Students who write the same number in the blank for two or more parts of question 4 need to understand that each expression is a name for 8 only when the correct number is placed in the $\square$ *and* that the correct number depends on the expression. For example, the expression $14 - \square$ can be a name for 5 when 9 is placed in the $\square$; in contrast, $14 - \square$ is also a name for 12 when 2 is placed in the $\square$.

In questions 2–4, the □ represents a particular number for each expression. The number that it represents depends on the number that is being named by the expression in which the □ is found.

Guiding Class Discussion

The practice sheets and suggested activities in the Follow-up Instruction section under the concept **Use of a variable as an unknown in an expression** may be helpful for those students having difficulty or needing more practice.

Focus of Question 5

Question 5 replaces the phrase "is a name for" by the equal sign in order to uncover several possible misconceptions concerning the equal sign.

Question 5

5. Consider each of the following. Is the statement true?

yes (no)	**a)**	$6 + 5 = 12$
(yes) no	**b)**	$10 = 12 - 2$
(yes) no	**c)**	$10 = 10$
(yes) no	**d)**	$1 + 9 = 5 + 5$
yes (no)	**e)**	$2 - 12 = 10$

Analysis of Question 5

Students who answer **no** to 5b may think that a single number is not permitted to the left of the equal sign. They may believe that the symbol "=" means that an answer to a problem always belongs to the right of "=". For example, students may think that $12 - 2 = 10$ is correct but that $10 = 12 - 2$ is not. Students who answer **no** to 5c may have difficulty accepting the fact that a number equals itself, while those who answer **no** to 5d may have difficulty with the idea that an operation on two numbers can be another name for the same operation on two different numbers. These ideas form an important foundation for the understanding of the role of the variable in an equation.

Guiding Class Discussion

The practice sheets and suggested activities in the Follow-up Instruction section under the concept **Use of a variable as a specific unknown in a linear equation** may be helpful for those students having difficulty or needing more practice.

Focus of Question 6

Question 6 asks the student to examine equations containing the symbol $\boxed{7}$ to ascertain their truth. This question is a lead-in to question 7, which asks the student to place the 7 in the placeholder.

Question 6

6. Consider each of the following. Is the statement true?

(yes)	no	**a)**	$\boxed{7} + 1 = 8$
(yes)	no	**b)**	$2 + \boxed{7} = 9$
(yes)	no	**c)**	$\boxed{7} = 5 + 2$
yes	(no)	**d)**	$\boxed{7} + 3 = \boxed{7} - 3$

Analysis of Question 6

Students who answer **yes** to 6d may believe the statement is true because the same symbol $\boxed{7}$ appears on both sides of the equation.

Guiding Class Discussion

The practice sheets and suggested activities in the Follow-up Instruction section under the concept **Use of a variable as a specific unknown in a linear equation** may be helpful for those students having difficulty or needing more practice.

Focus of Question 7

Question 7, which presents equations with the $\square$ as a placeholder, is an intermediate step to the use of letters as variables in equations. However, since the emphasis of this unit is on understanding the concept of variable and not on the technique of solving an equation, the value to be placed in the $\square$ is provided. The teacher should emphasize that, in this question, the only equations that are true are those in which 7 can be placed in the $\square$ to make a true statement.

Question 7

7. Consider each of the following. If 7 is placed in each $\square$, is the statement true?

(yes)	no	**a)**	$\square + 4 = 11$
(yes)	no	**b)**	$15 = 22 - \square$
(yes)	no	**c)**	$\square + 8 = 10 + 5$
yes	(no)	**d)**	$\square + 5 = \square + 6$
(yes)	no	**e)**	$\square + \square = 14$

Analysis of Question 7

In 7a, 7b, and 7c, the uses of the equal sign parallel 5a, 5b, and 5d, respectively. If the student incorrectly responds to 7a, 7b, or 7c, the teacher should consult the appropriate part of the discussion for question 5 to determine why the student may have misunderstood the question. The idea of using the same replacement value more than once in an equation occurs in 7d and 7e. A **no** answer to 7e may mean that the student placed 7 in one $\square$ only, yielding $7 = 14$, or that 7 was placed in one □ and some other number was placed in the second □ , yielding a result that was greater than or less than 14. In either case, it should be explained that □ + □ means that the same number is placed in both □ 's.

Students who answer **yes** to 7d should be asked whether $7 + 5$ and $7 + 6$ name the same number.

Guiding Class Discussion

To extend the idea, the teacher may want to ask if there is another way, using □ , to represent □ + □ . The discussion could involve asking students to find a number such that when that number is placed in the □ , the expression □ + □ will name 20. Then repeat the question again for numbers that will name 32 and 150. The discussion should lead to the fact that □ + □ is the same as $2 \cdot □$.

The practice sheets and suggested activities in the Follow-up Instruction section under the concept **Use of a variable as a specific unknown in a linear equation** may be helpful for those students having difficulty or needing more practice. **Practice sheet 3**, in particular, emphasizes the use of □ and letters as placeholders. It could be used to summarize the ideas presented in questions 5–7.

Focus of Questions 8–11

Questions 8–11 make the transition to the use of letters to represent variables and provide practice in substituting values in order to name a specific number.

Questions 8–11

8. Consider each of the following expressions. Is it a name for 6?

⟨yes⟩ no	**a)**	$3 + 3$
⟨yes⟩ no	**b)**	$3 + □$ when 3 is placed in the □
⟨yes⟩ no	**c)**	$3 + x$ when x is replaced by 3

9. Consider each of the following expressions. Is it a name for 9?

yes ⟨no⟩	**a)**	$3 + 3$
⟨yes⟩ no	**b)**	$32 - 23$
⟨yes⟩ no	**c)**	$□ + 8$ when 1 is placed in the □
yes ⟨no⟩	**d)**	$13 - □$ when 6 is placed in the □
⟨yes⟩ no	**e)**	$9 + c$ when c is replaced by 0
⟨yes⟩ no	**f)**	$28 - k$ when k is replaced by 19

10. Consider each of the following numbers. If g is replaced by the number, will $g + 21$ be a name for 34?

yes ⟨no⟩	**a)**	12
⟨yes⟩ no	**b)**	13
yes ⟨no⟩	**c)**	21
yes ⟨no⟩	**d)**	34

11. If each expression below is a name for 16, what number should replace y?

    a) In $y + 4$, _12_ should replace y.

    b) In $\dfrac{48}{y}$, _3_ should replace y.

    c) In $y - 4$, _20_ should replace y.

    d) In $\dfrac{y}{2}$, _32_ should replace y.

    e) In $50 - y$, _34_ should replace y.

Analysis of Questions 8–11

In question 8 the student must determine whether each of the three expressions is a name for the number 6. The teacher should be certain that the student is aware of the distinction between "is placed in" in 8b and "is replaced by" in 8c. While 8b is a transition to substitution, 8c involves the actual substitution of a number for a letter, a situation that the student will encounter repeatedly in the study of algebra.

Guiding Class Discussion

In question 10 the teacher should stress that when $g+21$ names a specific number, there is only one possible replacement value for g.

In question 11, as in questions 9e, 9f, and all parts of 10, the teacher should highlight the use of substitution as a technique to determine the replacement values. For example, in 11a, the teacher could ask the student to choose a number and add 4 to that number. The student could then be asked whether the result is equal to 16 and if it is not, whether the next number to be chosen should be greater than or less than the number that was chosen previously. Although the idea of inverse operations enters into this question, the focus should be on the substitution that is involved since this unit is emphasizing the role of variables in expressions, equations, and inequalities. The teacher should have the students determine the answer in 11a by repeated trial substitutions for y until a numerical expression that names the number 16 is obtained. The teacher should be certain that the student understands that numbers are to be *substituted* for the letters. If many students have difficulty with question 11, the teacher may want to construct a similar problem with letters and smaller numbers.

The practice sheets and suggested activities in the Follow-up Instruction section under the concept **Use of a variable as an unknown in an expression** may be helpful for those students having difficulty or needing more practice.

Focus of Questions 12–14

The purpose of question 12 is to reinforce the use of the substitution technique with expressions involving a letter. This is the first question in the assessment where the operation of multiplication is performed after the substitution has been made.

Questions 13–14 assess understanding of a concept that is often confusing for many students — the idea of concatenation with respect to the concept of variable, *i.e.*, the convention that a number followed immediately by a letter indicates the product of the number and the letter.

Questions 12–14

12. Consider each of the following expressions. Is it a name for 24?

(yes)	no	**a)**	$3 \cdot 8$
(yes)	no	**b)**	$4 \cdot 6$
(yes)	no	**c)**	$2 \cdot \square$ when 12 is placed in the $\square$
yes	(no)	**d)**	$6 \cdot \square$ when 8 is placed in the $\square$
(yes)	no	**e)**	$12 \cdot w$ when w is replaced by 2
yes	(no)	**f)**	$2 \cdot t$ when t is replaced by 4

13. $8 \cdot n$ can also be written as $8n$. Consider each of the following. If n is replaced by 5, is the statement true?

(yes)	no	**a)**	$8n$ is a name for $8 \cdot 5$.
yes	(no)	**b)**	$8n$ is a name for 85.
yes	(no)	**c)**	$8n$ is a name for $8 + 5$.
(yes)	no	**d)**	$8n$ is a name for 40.

14. $3c$ is a name for 36. Consider each of the following. Is the statement true?

yes	(no)	**a)**	c can be replaced by 6.
yes	(no)	**b)**	c can be replaced by 9.
(yes)	no	**c)**	c can be replaced by 12.
yes	(no)	**d)**	c can be replaced by 18.

Analysis of Questions 12–14

Question 12 begins by presenting several names for the number 24. The raised dot is used in question 12 to indicate multiplication; the convention of concatenation is not introduced until questions 13 and 14.

Question 13 presents an important convention in algebra that is confusing to students and is often not well emphasized in textbooks: a number followed immediately by a letter indicates the product of the number and letter. The teacher should reinforce the convention of concatenation as multiplication whenever the opportunity presents itself. Presenting examples in which students use the context of the problem to determine the exact role of the letters in that problem may also help clarify the convention. For example, $8m$ means 8 times m in algebra, but students may think that it represents 8 meters.

Two of the four parts of question 13 deal with very common misconceptions about $8n$ when n is replaced by 5. Students who answer yes to 13b do not understand the convention that $8 \cdot n$ can be written as $8n$. It is important that the student realize that substituting 5 for n in $8 \cdot n$ is the same as substituting 5 for n in $8n$. It is easier for the student to see that $8 \cdot n$ leads directly to $8 \cdot 5$, since the operation of multiplication is indicated by the raised dot. If a student does not make the link between $8n$ and $8 \cdot n$, he or she may incorrectly interpret $8n$ to be 85. Students who answer yes to 13c are not correctly connecting the concatenated form, $8n$, with the operation of multiplication. These students should be reminded that concatenation in algebra always represents the operation of multiplication.

Question 14 continues the convention that is introduced in question 13. However, in 14 the student must determine the correct replacement values for the variable in the expression $3c$ when $3c$ is a name for 36. A student who answers yes to 14a does not understand that $3c$ means 3 times c. A student who answers yes to 14b may have added the digits 3 and 6 in the number 36, while a yes answer to 14d may mean that the student performed an incorrect multiplication or division.

Guiding Class Discussion

The practice sheets and suggested activities in the Follow-up Instruction section under the concept **Use of a variable as an unknown in an expression** may be helpful for those students having difficulty or needing more practice. In particular, **practice sheet 4** emphasizes the concatenation convention and substitution to show equivalence of expressions using a word problem context.

Focus of Question 15

Question 15 encompasses both addition and multiplication expressions in equations that involve the use of letters as variables. In 15a and 15d, the convention of concatenation is reinforced.

Question 15

15. Consider each of the following. If y is replaced by 6, is the statement true?

yes	(no)	**a)**	$5y = 56$
yes	(no)	**b)**	$8 + y = 12$
(yes)	no	**c)**	$13 = 19 - y$
(yes)	no	**d)**	$42 = 7y$

Analysis of Question 15

It is possible that students who answer yes to 15a and no to 15d are still confused about the meaning of expressions like $5y$ or $7y$. In 15a, it should be emphasized that when 6 is substituted for y, $5y$ becomes $5 \cdot 6$ (not 56), while in 15d, when the same number is substituted for y, the result is $7 \cdot 6$ (not 76). In 15c, the arithmetic operation occurs to the right of the equal sign. The teacher should refer to the discussion of question 5, if necessary, for information regarding incorrect responses to this question.

Guiding Class Discussion

The practice sheets and suggested activities in the Follow-up Instruction section under the concept **Use of a variable as a specific unknown in a linear equation** may be helpful for those students having difficulty or needing more practice.

Focus of Questions 16–19

Questions 16–19 involve the use of a variable in an inequality. While a variable is used to represent a specific unknown in an equation, a variable in an inequality can represent several values. The development of the use of variables in inequalities parallels this unit's development of the use of variables in equations. The teacher should emphasize that there may be many possible replacement values for a variable in an inequality, as opposed to the one allowed replacement value for a variable in a linear equation.

Questions 16–19

16. Consider each of the following. Is the statement true?
(Remember that "$\square > \triangle$" means "$\square$ is greater than $\triangle$.")

(yes)	no	**a)**	$8 + 7 > 11$
yes	(no)	**b)**	$2 \div 24 > 11$
(yes)	no	**c)**	$11 > 7 + 2$
yes	(no)	**d)**	$11 > 7 + 6$
yes	(no)	**e)**	$11 > 7 + 4$

17. Consider each of the following numbers. If the number is placed in the $\square$, will $\square + 6$ name a number greater than 13?

yes	(no)	**a)**	5
yes	(no)	**b)**	6
yes	(no)	**c)**	7
(yes)	no	**d)**	8
(yes)	no	**e)**	9

18. Answer each of the following questions about $17 + \square > 35$.

yes	(no)	**a)**	Is $17 + \square > 35$ true if the number 18 is placed in the $\square$?
(yes)	no	**b)**	Is $17 + \square > 35$ true if any number greater than 18 is placed in the $\square$?
yes	(no)	**c)**	Is $17 + \square > 35$ true if any number less than 18 is placed in the $\square$?

19. Answer each of the following questions about $25 - z < 12$.
(Remember that "$\square < \triangle$" means "$\square$ is less than $\triangle$.")

yes	(no)	**a)**	Is $25 - z < 12$ true if the number 13 replaces z?
(yes)	no	**b)**	Is $25 - z < 12$ true if any number greater than 13 replaces z?
yes	(no)	**c)**	Is $25 - z < 12$ true if any number less than 13 replaces z?

Analysis of Questions 16–19

In question 16, the student must determine whether each numerical inequality is true. The meaning of the inequality symbol is supplied so that the student's response will depend on the understanding of the concept of "is greater than" rather than whether the student remembers the meaning of the symbol itself. At this time the teacher may want to discuss the fact that the inequality sign can be read left to right or right to left but that the interpretation depends on the direction in which it is read. Students who answer **yes** to 16b need to realize that $2 \div 24$ is not the same as $24 \div 2$. In 16c, 16d, and 16e, the operation is on the right side of the inequality. Teachers need to question students who answer **yes** to 16d to determine the source of the misunderstanding. The student may have determined the validity of the inequalities $11 > 7$ and $11 > 6$ separately, rather than adding 7 and 6 first, and then analyzing the inequality $11 > 13$. The student should understand that it is necessary to perform an indicated operation *first* before the validity of an inequality can be established.

Question 17 involves the use of the $\square$ as a placeholder in an inequality. The teacher should emphasize the idea that, because the expression $\square + 6$ names a number *greater than* 13, there is more than one number that can be placed in the $\square$. Also of importance in 17c is the fact that $\square + 6$ does *not* name a number *greater than* 13 when 7 is placed in the $\square$. Question 17c provides the opportunity for discussion of terms like "is equal to," "is less than," and "is greater than." However, it is best not to consider the terms "is greater than or equal to" and "is less than or equal to" at this point since it may confuse the student's understanding of the role of the variable in an inequality.

Question 18 may be difficult for students because it assesses whether the student can determine the set of values that should be placed in the $\square$ in the inequality $17 + \square > 35$. A **yes** answer to 18a may indicate that the student is viewing the inequality as an equation while a **yes** answer to 18c may mean that the student does not understand the meaning of the concept 'is greater than." If the student responds that only numbers greater than 18 can be placed in the $\square$, the teacher should still question further to be certain that the student realizes that many numbers exist that can be placed in the $\square$, and not just a single number that is greater than 18.

Question 19 extends the ideas presented in question 18 by asking the student to work with the "is less than" inequality and to use letters instead of $\square$'s, to represent variables. Since 19a, b, and c parallel 18a, b, and c, respectively, the teacher can refer to the appropriate parts of the discussion of question 18 if students have difficulty with question 19.

The teacher should evaluate carefully the consistency of a student's responses to 17, 18, and 19. The student may correctly respond that the replacement values must be greater than 18 and greater than 13 in questions 18 and 19, respectively, and still think that there is only one correct number that can be placed in the $\square$ in question 17. This inconsistency may indicate a lack of understanding of the fact that, in an inequality, a variable can have many values.

Guiding Class Discussion

Writing an inequality containing a letter for the variable on the chalkboard and surveying the class to determine various replacement values for the variable will help to reinforce the idea that a variable in an inequality can have more than one value.

The practice sheets and suggested activities in the Follow-up Instruction section under the concept **Use of a variable in a linear inequality** may be helpful for those students having difficulty or needing more practice. **Practice sheet 5**, which would be particularly useful at this point, is discussed under the aforementioned concept.

Focus of Questions 20–23

Questions 20–23 present the idea of a variable as a pattern generalizer, using tables to analyze the relationship between two quantities.

Question 21 assesses whether the students can identify a verbal description of a rule and whether students are analyzing the data vertically instead of horizontally.

Questions 22 and 23 involve the use of a variable to generalize the pattern in the table.

Questions 20–23

20.

Column A	Column B
1	6
2	7
3	8
4	9
5	*10*
12	*17*

In the table above, a rule is applied to each number in column A to get the corresponding number in column B. Determine the rule and use it to fill in the blanks in the table.

21.

Column A	Column B
1	7
2	14
3	21
4	28

Consider each of the following rules. Can the rule be applied to each number in Column A in the table above to get the corresponding number in column B?

yes (no)   **a)**   Add 6 to each number in column A.

(yes) no   **b)**   Multiply each number in column A by 7.

yes (no)   c)   Multiply each number in column A by $\frac{1}{7}$.

yes (no)   d)   Add 7 to each number in column A.

22. In the table below, a rule is applied to each number in the left column to get the corresponding number in the right column.

Number of Weeks	Number of Days
1	7
2	14
3	21
4	28

Consider each of the following expressions. If □ stands for the number of weeks and a number is placed in the □, will the expression give the corresponding number of days?

yes (no)   **a)** $\Box + 6$

(yes) no   **b)** $7 \cdot \Box$

yes (no)   **c)** $\frac{1}{7} \cdot \Box$

yes (no)   **d)** $7 + \Box$

23. In the table below, a rule is applied to each number in the left column to get the corresponding number in the right column.

Number of Weeks	Number of Days
1	7
2	14
3	21
4	28

Consider each of the following expressions. If n stands for the number of weeks and n is replaced by a number, will the expression give the corresponding number of days?

yes (no)   **a)** $n + 6$

(yes) no   **b)** $7n$

yes (no)   **c)** $\frac{1}{7}n$

yes (no)   **d)** $7 + n$

Analysis of Questions 20–23

In question 20, the students should analyze horizontally the data in the table to determine the relationship between columns A and B and then supply the two missing numbers in the table. The teacher should determine from various students in the class what rule they used to complete the table.

Students who answer **yes** to question 21a may be using only the first number pair in the table, while those who answer **yes** to 21c are applying the inverse rule or mixing the columns. A **yes** answer to 21d may indicate that the student analyzed vertically the numbers in column B to determine the pattern.

In question 22, the □ is used in place of the number of weeks and in 23, a letter represents the number of weeks. The days/weeks setting was introduced in question 22 to give the students a concrete example of a pattern generalizer. In question 23, the convention of using $7n$ to represent $7 \cdot n$ should again be reinforced. Students who miss question 23 but not question 22 are likely to be having trouble using letters to represent an unconstrained variable.

Guiding Class Discussion

The practice sheets and suggested activities in the Follow-up Instruction section under the concept **Use of a variable as a pattern generalizer** may be helpful for those students having difficulty or needing more practice. **Practice sheet 6**, in particular, can be used for practice in using given data to determine, use, and write rules, in words and symbols.

Focus of Questions 24–25

Questions 24 and 25 illustrate the idea that variables can depend on one another. The teacher can discuss this idea with the students without using the terms "dependent variable" and "independent variable." It will be easier for the students to understand this concept if the teacher says that the value of one variable *depends on* the value of another variable when the two variables are in the same equation or inequality. After the students have answered the questions in the Instructional Assessment, the teacher should discuss with the students how they determined the numbers to write in the blanks. It should be stressed that, while the students can choose some numbers, other numbers are the result of computational procedures that are performed on previously chosen numbers.

Questions 24–25

For questions 24–25 fill in the blanks with numbers. Be sure the numbers agree with the facts.

24. Mary earns 3 dollars an hour.

    a)   On Tuesday Mary worked 5 hours and earned a total of ___*15*___ dollars for that day.

    b)   On Wednesday Mary worked _*varies*_ hours at 3 dollars an hour and earned a total of _*varies*_ dollars for that day. (Supply your own numbers.)

25. Anna earns 4 dollars an hour.

    a)   On Monday, she worked n hours. Use n to write an expression that shows the number of dollars that Anna earned.

        Answer: _*4n*_

b) On Thursday, the total number of dollars Anna earned was 4y. How many hours did Anna work?

Answer: _y_____

Analysis of Questions 24–25

In question 24, the number of dollars earned depends on the number of hours worked. Also, the number of dollars earned in an hour varies from setting to setting. In question 25, the convention of concatenation is related to a real-world setting. The teacher should question the students to be certain that they understand that the total amount of dollars earned is determined by multiplying the total number of hours worked times the number of dollars earned in one hour. If the student does not have this understanding of how multiplication relates to the variables in the problem, it will be very difficult to see that, from the expression 4y, one can determine that Anna worked y hours (in the second part of question 25). Also, in the first part of 25, students may have difficulty constructing an expression since specific numerical values are not provided. In an algebraic expression such as 4n, the operation of multiplication is *represented* implicitly by concatenation.

Guiding Class Discussion

The practice sheets and suggested activities in the Follow-up Instruction section under the concept **Use of a variable in formulas and word problems** may be helpful for those students having difficulty or needing more practice. In particular, **practice sheet 7**, discussed under this concept, can be used to emphasize that the value of one variable often depends on the value of another variable.

Focus of Question 26

Question 26 is the first in the Instructional Assessment that involves the use of two letters as variables in an equality situation. The teacher should point out that while *lw* is a name for the area of a rectangle, it also represents the operation *l* times *w*.

Question 26

26.

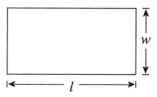

The area of the rectangle above can be found by multiplying its length by its width. If *l* stands for length and *w* stands for width, then *lw* is another name for the area of the rectangle.

Consider each of the following. Is it true?

yes ⃝ no **a)** If the length is 6 and the width is 3, then 6 · 3 is another name for *lw*.

yes ⃝no **b)** If the length is 6 and the width is 3, then 6 + 3 is another name for *lw*.

yes ⃝no **c)** If the length is 6 and the width is 3, then 63 is another name for *lw*.

yes ⃝ no **d)** If the length is 6 and the width is 3, then 18 is another name for *lw*.

Analysis of Question 26

Students who answer **no** to 26a and/or **yes** to 26c may need to be reminded that, in algebra, $l \cdot w$ is written *lw*. A **yes** answer to 26b may indicate that the student thinks $l + w$ is another name for *lw* when l and w are replaced by 6 and 3, respectively. Question 26d is an extension of 26a. Students who answer 26a correctly and 26d incorrectly, or vice versa, need to be questioned to determine the source of their misunderstanding, since these two statements are equivalent.

Guiding Class Discussion

The practice sheets and suggested activities in the Follow-up Instruction section under the concept **Use of a variable in formulas and word problems** may be helpful for those students having difficulty or needing more practice. **Practice sheet 8**, discussed under this concept, also gives the student additional practice in the other concepts developed in questions 1–26.

Focus of Questions 27–28

Questions 27 and 28 involve the use of two variables in an *inequality*.

Questions 27–28

Questions 27–28 refer to the following information.

The length of a certain rectangle is greater than the width. If l stands for the length and w stands for the width, then, in symbols, $l > w$.

27. Consider each pair of numbers in the table below. Is $l > w$ for that pair of numbers?

		l	w
yes ⃝no	**a)**	7	9
⃝yes no	**b)**	3	2
yes ⃝no	**c)**	5	15
⃝yes no	**d)**	13	10
⃝yes no	**e)**	13	12

28. If $l > w$, consider each of the following statements. Is the statement true?

   (yes) no   **a)**   If l is 7, then w is less than 7.

   (yes) no   **b)** l could be 15 when w is 10.

   (yes) no   **c)**   If w is 13, then l is greater than 13.

   yes (no)   **d)** l and w could have the same value.

Analysis of Questions 27–28

Question 27 asks the students to determine which pairs of numbers (l, w) will satisfy the inequality $l > w$. In this question, the teacher should focus the discussion on the technique of substitution to determine whether each proposed pair is valid. Also, since the question involves an inequality, the teacher should emphasize that there is a variety of numbers that will make the inequality a true statement. The latter point can be reinforced by showing the students that the pairs in both 27d and 27e satisfy the inequality. The students could then be asked for what other values of w will $l > w$ be true when l is 13.

In question 28, the student must evaluate several general statements regarding $l > w$. Questions 28a and 28c involve generalizing a description for a set of values of one variable in an inequality when the value of the other variable is fixed. (This addresses again the idea that the value of one variable might depend on the value of another variable.) In 28b, when the value of w is fixed, the student must determine if 15 could be a value for l. The teacher should ask the student if there are other values possible for l when $l > w$ and w is 10 to assess for understanding of the idea that a variable in an inequality can often assume more than one value. Question 28d addresses the fact that if $l > w$, it is not possible for l and w to be equal.

Guiding Class Discussion

The practice sheets and suggested activities in the Follow-up Instruction section under the concept **Use a variable in formulas and word problems** may be helpful for those students having difficulties or needing more practice.

Follow-up Instruction

After the Instructional Assessment has been used to pinpoint students' conceptual weaknesses, the teacher may wish to use the following suggested activities and practice sheets to help correct those misconceptions.

Use of a variable as an unknown in an expression

The purpose of **Practice sheet 1** is to help students to realize that a number can be expressed in many different ways by having them generate different forms of certain numbers. For example, the number 1,988 can be named by $2,000 - 12$ or $1,388 + 600$ or $2 \cdot 994$ or $5,964 \div 3$. To make the activity more interesting for the students, the teacher can also use numbers such as the school's address, or the student's address, or the current calendar year. The teacher may also want to limit each expression to a single operation; otherwise the rule of order of operations will become an additional concern.

Practice sheet 2 asks the students to express the numbers from 1 to 100 using only four digits, in order, and symbols for operations. Again, to make this activity more interesting to students, their expected year of graduation can be used. This activity will give the teacher a chance to discuss the order of operations if he or she desires. As the students learn more mathematics, they will have more operation symbols to use, *e.g.,* $\sqrt{}$. This activity can be used over a long period of time.

The teacher can further assess understanding of concatenation by using activities such as those found in **Practice sheet 4**. In each case, the focus should be first on understanding the role of the variable in the expression and then on the technique of substitution to determine a specific value.

Follow-up Assessment questions 1–4, 8, 9, and 10–14 also address this concept.

Use of a variable as a specific unknown in a linear equation

Practice sheet 3 reviews the ideas presented in assessment questions 5–7. Question 1 on the practice sheet checks for understanding about the equal sign. Question 2 introduces the symbol $\boxed{15}$. Since the same number appears more than once in several of the statements, questions

1 and 2 can be used to help students to understand the idea that the same number must be placed in each □ that occurs in a given equation.

Each part of question 3 has a slightly different purpose. Part a addresses the use of a variable as a specific unknown. In part b, the arithmetic operation is to the right of the equal sign. Parts c, d, and e all involve placing the 9 in more than one □ in the same equation. Part f involves the use of a combination of ideas presented in b through e.

Question 4 extends the idea of Question 3 by using a letter, instead of the □ , to represent a variable.

This practice sheet can serve as a review for students who have difficulty with question 15, particularly parts b and c.

Questions 5–7 and 15 in the Follow-up Assessment also relate to this concept.

Use of a variable in a linear inequality

Practice sheet 5 could be used when the teacher is discussing the role of a variable in inequalities such as $3 + n > 18$ or $y - 5 < 4$. The focus on the role of the variable in an inequality should first be on the fact that it can often assume a variety of values and then on the fact that it has a least or greatest value when the set of replacement values is the set of integers. Students often see the least or greatest value as the *only* possible value of the variable. The teacher needs to assure the student that the least or greatest value is a benchmark and that all values less than or greater than (depending on the particular inequality) the benchmark will also make the inequality valid. This idea can be reinforced by having the student graph some of the inequalities on a number line so that the solution set can be seen more clearly.

Note: Some students may argue that 24 is the correct answer for question 8 on practice sheet 5 because $24 \div 3 = 8$. In this case, the teacher should point out that it is the number in the box, not the value of $\frac{\square}{3}$, that must be a whole number or elicit this comment from the students themselves through discussion.

Follow-up Assessment questions 16–19 also deal with this concept.

Use of a variable as pattern generalizer

Practice sheet 6 gives two more examples of how tables can be used to represent the relationship between two quantities. The students are asked to determine the rule that relates the two columns, expressing it both in words and symbols, and to use the rule to calculate one number given the other number in the pair. If necessary, the teacher may wish to generate additional tables and put them on the blackboard for class discussion. Alternatively, students may create their own problems by starting with a rule, then using it to generate a table of several pairs of values; students could then exchange problems, returning the answers to the problem's creator for grading.

This concept is also addressed by questions 20–23 in the Follow-up Assessment.

Use of a variable in
formulas and word problems

Practice sheet 7 uses a word problem format to present situations in which the value of one variable depends on the value of another variable. The teacher may want to encourage students to work together and to discuss their solution procedures for this practice sheet.

Practice sheet 8 presents situations that use variables in a formula and a word problem. It also reviews ideas about different names for a given number, concatenation and other concepts presented in the unit.

Practice sheet 9 provides an opportunity for the student to consider inequalities, expressed in symbols, in real-life contexts. After completing this activity, the students could be asked to construct their own situations involving inequalities, expressing the situation first in words, then in symbols.

Questions 24–28 in the Follow-up Assessment also relate to this concept.

Bibliography of Resource Materials

Booth, L.R. *Algebra: Children's Strategies and Errors.* Windsor, U.K.: NFER Nelson, 1984.

Mason, J., A. Graham, D. Pimm, and N. Gowar. *Routes to/Roots of Algebra.* Milton Keynes, U.K.: The Open University Press, 1985.

Usiskin, Z. "Conceptions of School Algebra and Uses of Variables." *1988 Yearbook of the National Council of Teachers of Mathematics:* 8–19.

Answers

Answers to Instructional Assessment

1. a) yes
   b) yes
   c) yes
   d) yes
   e) no
   f) no
   g) yes
   h) yes
2. a) yes
   b) no
   c) yes
   d) yes
   e) no
3. a) no
   b) no
   c) yes
   d) no
   e) no
4. a) 5
   b) 11
   c) 64

   d) 18
   e) 2
5. a) no
   b) yes
   c) yes
   d) yes
   e) no
6. a) yes
   b) yes
   c) yes
   d) no
7. a) yes
   b) yes
   c) yes
   d) no
   e) yes
8. a) yes
   b) yes
   c) yes
9. a) no
   b) yes

   c) yes
   d) no
   e) yes
   f) yes
10. a) no
    b) yes
    c) no
    d) no
11. a) 12
    b) 3
    c) 20
    d) 32
    e) 34
12. a) yes
    b) yes
    c) yes
    d) no
    e) yes
    f) no
13. a) yes
    b) no

    c) no
    d) yes
14. a) no
    b) no
    c) yes
    d) no
15. a) no
    b) no
    c) yes
    d) yes
16. a) yes
    b) no
    c) yes
    d) no
    e) no
17. a) no
    b) no
    c) no
    d) yes
    e) yes
18. a) no

    b) yes
    c) no
19. a) no
    b) yes
    c) no
20. 10; 17
21. a) no
    b) yes
    c) no
    d) no
22. a) no
    b) yes
    c) no
    d) no
23. a) no
    b) yes
    c) no
    d) no
24. a) 15
    b) Answers
       will vary

    but the
    second
    number
    should be
    5 times
    the first.
25. a) $4n$
    b) y
26. a) yes
    b) no
    c) no
    d) yes
27. a) no
    b) yes
    c) no
    d) yes
    e) yes
28. a) yes
    b) yes
    c) yes
    d) no

Answers to Practice Sheets

Practice sheet 1

Answers will vary.

Practice sheet 2

Answers will vary.

Practice sheet 3

1. a) yes
   b) yes
   c) yes
   d) no
   e) yes
2. a) yes
   b) yes
   c) yes
   d) no
   e) no
   f) yes
3. a) yes
   b) yes
   c) no
   d) yes
   e) yes
   f) yes
4. a) no
   b) no
   c) yes
   d) yes
   e) yes
   f) yes

Practice sheet 4

1. $3 \times k$ or $3 \cdot k$ or $3k$
   42

2. $5 \times r$ or $5 \cdot r$ or $5r$

r	$r + r + r + r + r$	$5 \cdot r$	$5r$
1	$1 + 1 + 1 + 1 + 1$	$5 \cdot 1$	5
2	$2 + 2 + 2 + 2 + 2$	$5 \cdot 2$	10
3	$3 + 3 + 3 + 3 + 3$	$5 \cdot 3$	15
5	$5 + 5 + 5 + 5 + 5$	$5 \cdot 5$	25
12	$12 + 12 + 12 + 12 + 12$	$5 \cdot 12$	60
20	$20 + 20 + 20 + 20 + 20$	$5 \cdot 20$	100

Practice sheet 5

1. 15
2. 5

3. 58
4. 9
5. 2
6. 3
7. 64
8. 26

Practice sheet 6

1. a) 20, 26, 32
   b) Subtract 3 from the number in column A to get the corresponding number in column B.
   c) $w - 3$
2. a) 33, 61, 81
   b) Multiply the number in column A by 2 and add 1 to get the corresponding number in column B.
   c) $2v + 1$

Practice sheet 7

1. 43, 14
   $n, n/4$ (answers may vary)

Number of Chickens Sold	Number of Horses Sold
0	3
2	2
4	1
6	0

2. 10, 15
3. $8 + k$
4. $10 + k + 5$ or $15 + k$
5. 21

Practice sheet 8

1. a) yes
   b) yes
   c) yes
   d) no
   e) no
2. a) no
   b) yes
   c) no
   d) no
   e) yes

Practice sheet 9

1. a) yes

b) no
c) yes
d) no
e) no
2. a) yes
b) no
c) yes
d) no

e) no
3. Answers will vary. For example,

m	s
10	7, 8
12	9, 10
27, 16	15
8, 18	6
7	5, 2

4. $120.00

Answers to Follow-up Assessment

1. a) yes
b) yes
c) no
d) no
e) yes
f) yes
g) yes
h) yes
2. a) no
b) yes
c) yes
d) no
e) no
3. a) yes
b) no
c) no
d) no
e) no
4. a) 7
b) 17

c) 180
d) 14
e) 2
5. a) no
b) yes
c) yes
d) yes
e) yes
6. a) yes
b) no
c) no
d) no
7. a) yes
b) no
c) yes
d) yes
e) no
8. a) yes
b) yes
c) yes

9. a) no
b) no
c) no
d) yes
e) yes
f) no
10. a) no
b) no
c) yes
d) no
11. a) 2
b) 8
c) 10
d) 12
e) 44
12. a) yes
b) no
c) yes
d) no
e) no

f) yes
13. a) no
b) yes
c) yes
d) no
14. a) no
b) yes
c) no
d) no
15. a) yes
b) no
c) yes
d) yes
16. a) yes
b) yes
c) no
d) no
e) no
17. a) no
b) no

c) no
d) yes
e) yes
18. a) no
b) yes
c) no
19. a) no
b) no
c) yes
20. 108, 180
21. a) yes
b) no
c) no
d) no
22. a) no
b) no
c) yes
d) no
23. a) yes
b) no

c) no
d) no
24. a) 20
b) Answers will vary.
25. a) $6n$
b) z
26. a) yes
b) no
c) no
d) yes
27. a) yes
b) no
c) yes
d) yes
e) no
28. a) yes
b) yes
c) yes
d) yes

ALGE**BRIDGE**™

Concept of

Equality and

Inequality

Concept of Equality and Inequality

Introduction

One concept in algebra that students often find difficult to grasp is the meaning of the equal sign in an algebraic equation. For most students, early experiences in mathematics are based primarily on viewing the equal sign as a signal to perform an operation and give the answer (*e.g.*, $3 \times 2 = $) rather than a relation. A poor understanding of the relationship between the left and right sides of an equation may be one source of difficulty in solving the equation.

For example, in the equation $38 + 52 = 90$ students read from left to right and observe that the equal sign marks the addition of 38 and 52 and indicates the result. However, students may not perceive the role of the equal sign in a decomposition, as in the equation $8x = 3x + 5x$, or as an equivalence, as in the equation $3y + 18 = 4y + 10$. A decomposition has one term on the left side of the equal sign and several terms on the right side. In $8x = 3x + 5x$, the term $8x$ is decomposed into the terms $3x$ and $5x$. In an equivalence, both sides are numerically equal for some value of the unknown. For example, in the equation $3y + 18 = 4y + 10$, $3y + 18$ is equivalent to $4y + 10$ when $y = 8$.

In the context of simple linear equations and inequalities that involve whole numbers, this unit develops the student's ability to

-   understand the concept of equality;

-   understand the role of the equal sign

    in a decomposition;

    in an equivalence;

-   understand the concept of inequality;

-   understand the similarities and differences between equations and inequalities. For example,

    inequalities have multiple solutions;

    every inequality has an associated equation and a complementary inequality;

    both equations and inequalities have meaning when the left and right sides are interchanged, but, in an inequality, interchanging the sides changes the meaning;

-   construct numerical inequalities.

The teacher should have at least one scale with a set of weights in the classroom, since the scale provides an excellent physical model for the concept of equality. Teachers often speak of equations being balanced, and the use of a double-pan scale to model some equations reinforces the fact that one side of an equation is equal to the other side. Many of the primary concepts of this unit depend on the use of a scale to model them.

Cross-References to to Pre-Algebra and Algebra Textbook Topics

What follows is a compilation of chapter titles and index entries that are found in typical pre-algebra and algebra textbooks. The concepts and abilities addressed in this unit could be used to enhance the instruction on these topics.

Note: Only nonnegative numbers are used in this unit.

ratio and proportion   **formulas (geometry)**
number properties and equations **functions and graphs**
integers and equations   **linear equations**
rational numbers and equations **linear inequalities**
graphing equations and inequalities **variation**

List of Terms

The following is a list of words and phrases that are found in this unit. The teacher may want to add other words and phrases to this list.

balance   **equation**   **is not the same as**
can be written as **inequality**   **is the same as**
condition   **is greater than** **number sentence**
equals   **is less than**   **replacement set**

Classification by Concept

Target Concepts and Abilities	Questions in Instructional Assessment	Questions in Follow-up Assessment	Practice Sheets
Understand the concept of equality	1, 2	1, 2, 3, 4	1, 2
Understand the role of the equal sign in a decomposition	3, 4, 5, 6, 14	5, 6, 7, 8, 9	1
Understand the role of the equal sign in an equivalence	7, 8, 15, 17, 28, 29	10, 11, 12, 13	2
Understand the concept of inequality	9, 10, 11, 12, 13	14, 15, 16, 17, 18	3, 4, 5
Understand the similarities and differences between equations and inequalities	16, 18, 19, 20, 21, 22, 23, 24, 25, 26, 27	19, 20, 21, 22, 23, 24, 25, 26, 27, 28	3, 4
Construct numerical inequalities	30, 31	29, 30	5

Classification by Question

Target Concepts and Abilities		Understand the concept of equality	Understand role of equal sign in a decomposition	Understand role of equal sign in an equivalence	Understand concept of inequality	Understand similarities and differences between equations and inequalities	Construct numerical inequalities
Question Number or Practice Sheet Number	1	IA FA PS	PS				
	2	IA FA PS		PS			
	3	FA	IA		PS	PS	
	4	FA	IA		PS	PS	
	5		IA FA		PS		PS
	6		IA FA				
	7		FA	IA			
	8		FA	IA			
	9		FA		IA		
	10			FA	IA		
	11			FA	IA		
	12			FA	IA		
	13			FA	IA		
	14		IA		FA		
	15			IA	FA		
	16				FA	IA	
	17			IA	FA		

IA - Questions in Instructional Assessment
FA - Questions in Follow-up Assessment
PS - Practice Sheets

Classification by Question

Target Concepts and Abilities	Understand the concept of equality	Understand role of equal sign in a de-composition	Understand role of equal sign in an equivalence	Understand concept of inequality	Understand similarities and differences between equations and inequalities	Construct numerical inequalities
Question Number or Practice Sheet Number 18				FA	IA	
19					IA FA	
20					IA FA	
21					IA FA	
22					IA FA	
23					IA FA	
24					IA FA	
25					IA FA	
26					IA FA	
27					IA FA	
28			IA		FA	
29			IA			FA
30						IA FA
31						IA

IA - Questions in Instructional Assessment
FA - Questions in Follow-up Assessment
PS - Practice Sheets

Instructional Assessment

The questions in both the Instructional and Follow-up Assessment Instruments for this unit have different formats appropriate to the nature of the concepts addressed in the questions.

It is necessary to introduce students to these formats since they may have limited experience with different types of responses.

If there is not sufficient time to review examples of the different question types, review the directions below with the students before they work the *Algebridge*™ exercises.

Directions

There are two types of questions in this exercise.

The first type gives you several choices related to a specific question. You must answer **yes** or **no** for each of these choices.

The second type asks you to write your own answer to the problem.

Focus of Question 1

Question 1 assesses whether students can recognize an equation and understand the concept of equality.

Question 1

1. Consider each of the following. Is it an equation?

yes (no)	**a)**	$2 + 5$
yes (no)	**b)**	$2 + 5 =$
(yes) no	**c)**	$2 + 5 = 7$
yes (no)	**d)**	$7 + \ = 5$
yes (no)	**e)**	$2 + 5 > 6$
(yes) no	**f)**	$2 \times 5 = 9 + 1$
(yes) no	**g)**	$12 = 5 + 7$
(yes) no	**h)**	$2 \times 3 = 3 + 2$

225

Analysis of Question 1

An equation, as defined in this unit, is a mathematical statement that contains two expressions with an equal sign between them. Students who answer **yes** to 1a should be reminded that $2 + 5$ is an example of a mathematical expression, while those who answer **yes** to 1b may be anticipating the answer for the problem $2 + 5$ because they are viewing $=$ as an operational symbol in this statement. It should be pointed out to the students that 1d has an incomplete left side; *i.e.*, $7 +$ is not a complete expression since there is no term to the right of $+$. The statements in 1f and 1g illustrate the role of the equal sign in an equivalence and in a decomposition, respectively. Some students may believe that 1f is incorrect because it does not have an "answer" on the right side as in 1c. In this case, it should be explained that since 10 is the result of each of the two expressions 2×5 and $9 + 1$, the statement $2 \times 5 = 9 + 1$ is a true equation. The left and right sides are equal. Some students may believe that the equation $12 = 5 + 7$ (1g) is written backwards. Since they read naturally from left to right and know that the result of adding 5 and 7 is 12, the statement $5 + 7 = 12$ makes sense to them but $12 = 5 + 7$ does not. The students should understand that the statement is saying that one way to break down (or decompose) 12 is $5 + 7$. Some students may say that 1h is not an equation because it is false. They should be reminded that in this unit an equation is defined as a mathematical statement that contains two expressions with an equal sign between them. Both true and false equations are allowed under this definition.

Guiding Class Discussion

The practice sheets and suggested activities in the Follow-up Instruction section under the concept **Understand the concept of equality** may be helpful for those students having difficulty or needing more practice.

Focus of Question 2

Question 2 assesses the same concepts as question 1, but the equations in question 2 involve the $\square$ and literal terms.

Question 2

2. Consider each of the following. Is it an equation?

(yes)	no	**a)**	$3 + \square = 17$
yes	(no)	**b)**	$\square + 12$
yes	(no)	**c)**	$41 + \quad = \square$
yes	(no)	**d)**	$2 + \square$
(yes)	no	**e)**	$14 = 7 + \square$
(yes)	no	**f)**	$3 \times \square = 4 \times 5$
yes	(no)	**g)**	$5 + \square > 1$
(yes)	no	**h)**	$2 \times \square = 13 + 9$
(yes)	no	**i)**	$\square = 12$
yes	(no)	**j)**	$y - 9$
(yes)	no	**k)**	$x + 48 = 30 + 2x$

Analysis of Question 2

Since some students believe that equations contain numbers only, the teacher should emphasize that 2a, 2e, 2f, 2h, 2i, and 2k are all well-formed equations. However, whether the equations are true depends on which number is placed in the $\square$ or replaces the x or y in each equation. Students who answer **yes** to 2b or 2j need to realize that these are examples of mathematical expressions. Question 2e is an example of a decomposition and 2f, 2h, and 2k are examples of equivalences. Students who answer **yes** to 2c should be shown that this statement is not well-formed.

Guiding Class Discussion

The following discussion may help students understand equations that contain unknowns. The teacher should begin by writing a true equation on the chalkboard; for example $3 \times 8 + 4 = 28$. Then, a $\square$ is drawn around one of the numbers in the equation. One resulting equation could be $3 \times \boxed{8} + 4 = 28$. (Other possibilities are $\square \times 8 + 4 = 28$ and $3 \times 8 + \square = 28$.) If a student is now asked what number should be placed in the $\square$ to make the equation $3 \times \square + 4 = 28$ true, the answer 8 should be apparent from the previous sequence of steps. If this exercise is repeated with several different equations, the students should be able to view the $\square$ as a missing number in the equation. The order of steps in this exercise is significant because it helps the students to establish an understanding of equations that contain unknowns by building on the students' inherent understanding of the associated numerical equations. Once this link has been made, the teacher can make the transition from the $\square$ to the use of a literal term as a variable, explaining that, like the $\square$, the letter is used to "hide" a number in the equation.

The practice sheets and suggested activities in the Follow-up Instruction section under the concept **Understand the concept of equality** may be helpful for those students having difficulty or needing more practice.

Focus of Question 3

The focus of question 3 is on the concept of equality as a decomposition and on different words and phrases that mean the same as the equal sign.

Question 3

3. Consider each of the following. Is the statement another way to say "$20 + 16 = 36$"?

(yes)	no	**a)**	$36 = 20 + 16$
(yes)	no	**b)**	$20 + 16$ equals 36.
(yes)	no	**c)**	$20 + 16$ is the same as 36.
yes	(no)	**d)**	$20 + 16 > 36$
(yes)	no	**e)**	36 can be written as $20 + 16$.
(yes)	no	**f)**	36 is $20 + 16$
yes	(no)	**g)**	$20 + 16$ is greater than 36.

yes (no)   **h)**   36 is not the same as 20 + 16.
(yes) no   **i)**   36 is a name for 20 + 16.

Analysis of Question 3

Students who answer **no** to 3a and 3f may think that these statements are written backwards since the statements are read from left to right. In 3a they may view the problem as 20 + 16 and the answer or result as 36. It is important that students realize that the left and right sides of an equation can be interchanged and the equality will be maintained. This is one reason why 36 = 20 + 16 is the same as 20 + 16 = 36. Students should be encouraged to interpret 36 = 20 + 16 as the number 36 being broken down into 20 and 16 or 36 being expressed as 20 + 16.

Different words and phrases that mean the same as the equal sign are presented in 3b, 3c, 3e, 3f, and 3i. Emphasizing to students that these are all interpretations of = will help the students when they are translating word problems into equations in algebra.

Some students who answer **no** to 3a, 3e, 3f, and 3i may think that "another way to say" means that the order of terms must be preserved. Their answers should not be marked wrong as long as they understand that 3a, 3e, 3f, and 3i are alternate forms of the original statement.

Guiding Class Discussion

The practice sheets and suggested activities in the Follow-up Instruction section under the concept **Understand the role of the equal sign in a decomposition** may be helpful for those students having difficulty or needing more practice.

Focus of Questions 4–5

Questions 4 and 5 assess whether students can form a true equation by moving the numbers and symbols into different positions. Question 4 also involves an application of the commutative property of addition.

Questions 4–5

Questions 4–5 refer to the following information.

$$16 + 20 = 36$$

The arrangement of tiles shown above is one way to write a true equation that uses the tiles 36, 20, 16, $+$, and $=$.

4. What is a different arrangement that uses these same five tiles to form a true equation that begins with 20?

Answer: $20 + 16 = 36$

5. What is a different arrangement that uses the same tiles to form a true equation that begins with 36?

Answer:

| 36 | = | 20 | + | 16 |

Analysis of Questions 4–5

In question 4 the students should realize that $16 + 20$ and $20 + 16$ both add up to the same number. Question 5 involves the concept of decomposition. The fact that the decomposition is tied to the previous equation, $16 + 20 = 36$, should help the students better understand this concept.

Guiding Class Discussion

The practice sheets and suggested activities in the Follow-up Instruction section under the concept **Understand the role of the equal sign in a decomposition** may be helpful for those students having difficulty or needing more practice.

Focus of Question 6

Question 6 presents the concept of decomposition using a double-pan scale.

Question 6

6.

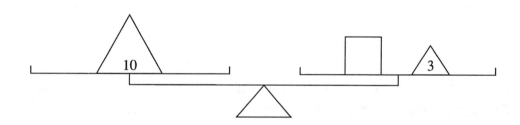

In the figure above, the weight on the left pan of the scale balances the weights on the right pan. The □ in this figure stands for an unknown weight. Consider each of the following. Could the statement describe the figure?

(yes)	no	**a)**	$10 = 7 + 3$
(yes)	no	**b)**	$10 = □ + 3$, when 7 is placed in the □.
yes	(no)	**c)**	$10 = 5 + 5$
yes	(no)	**d)**	$10 = □$

Analysis of Question 6

The number sentences in 6a and 6b could both describe the figure. Students who answer **no** to 6b should be questioned to see if they understand that when 7 is placed in the □, the equation will be true (*i.e.*, both sides will balance). Students who answer **yes** to 6c should be reminded that, while $10 = 5 + 5$ is a true equation, it does not describe the figure. In 6d, □ cannot be equal to 10 since the right pan of the scale in the figure also contains a 3 (*i.e.*,, the scale does not balance when 10 replaces □).

Guiding Class Discussion

The practice sheets and suggested activities in the Follow-up Instruction section under the concept **Understand the role of the equal sign in a decomposition** may be helpful for those students having difficulty or needing more practice.

In particular, **practice sheet 1** builds on the ideas about decomposition given in questions 3 and 6 by presenting decompositions of equations involving two variables.

Focus of Questions 7-8

An understanding of the equivalence of the left and right sides of an equation is assessed in questions 7 and 8.

Questions 7-8

Questions 7-8 refer to the following information.

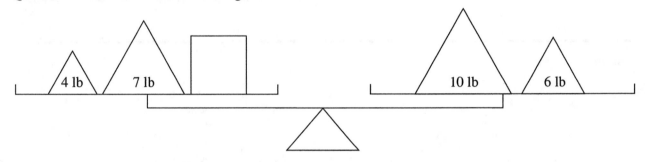

In the figure above, the weights on the left pan of the scale balance the weights on the right pan. The □ in this figure stands for an unknown weight. (lbs means pounds.)

7. Write a number sentence that uses □ and = to describe the figure.

    Answer: *For example,* $4 + 7 + □ = 10 + 6$

8. What is the weight of the □, in pounds?

    Answer: 5

Analysis of Questions 7-8

In question 7 the students are asked to write a number sentence that corresponds to the figure. The teacher should question those students who combine numbers on each side of their equation, *e.g.*, $11+□ = 16$, as well as those students who solve for □ and get $□ = 5$, to determine if they understand the concept of equality as presented in the figure. In question 8 the students are asked to determine what number goes in the □. This should be done by using a knowledge of number facts and not by using algebraic methods to solve the equation.

Guiding Class Discussion	The practice sheets and suggested activities in the Follow-up Instruction section under the concept **Understand the role of the equal sign in an equivalence** may be helpful for those students having difficulty or needing more practice.
Focus of Question 9	Question 9 assesses the students' understanding of the concept of inequality and their ability to recognize an inequality. In this unit an inequality is a mathematical statement that contains two expressions with $>$ or $<$ between them. (To emphasize understanding of the basic concept of inequality, only strict inequalities will be addressed in this unit.)

Question 9

9. Consider each of the following. Is it an inequality? (Remember that "$\square < \triangle$" means "$\square$ is less than $\triangle$" and "$\square > \triangle$" means "$\square$ is greater than $\triangle$.")

yes	(no)	**a)**	$5 - 2$
yes	(no)	**b)**	$2 + 1 >$
(yes)	no	**c)**	$3 + 4 > 5$
(yes)	no	**d)**	$2 + 4 < 3 + 4$
yes	(no)	**e)**	$3 > 2 +$
(yes)	no	**f)**	$5 + 2 < 6$

Analysis of Question 9	The symbols in 9a form an expression, not an inequality, and the statements in 9b and 9e are not well-formed inequalities. The teacher should point out to the students that inequalities, like equations, have a left side and a right side. The statement in 9b has no right side and the statement in 9e has an incomplete right side. While the statement in 9f fits the definition of an inequality, it is not a true inequality since 7 is not less than 6. Students may need to be questioned to see if they understand that an inequality may be true or false.
Guiding Class Discussion	In the discussion of this question, it is important that the teacher point out the difference between an equation and an inequality. In a true equation, the two sides are equal to the same numerical value but in a true inequality, the numerical value of one side is greater than the numerical value of the other side. The teacher may also want to discuss with the students ways to remember which symbol means "is greater than" and which means "is less than."
	The practice sheets and suggested activities in the Follow-up Instruction section under the concept **Understand the concept of inequality** may be helpful for those students having difficulty or needing more practice.

Focus of Question 10   Question 10 continues assessment of the same concepts as question 9, but with unknowns being represented by the □ and by literal terms.

Question 10

10. Consider each of the following. Is it an inequality?

yes (no)	**a)**	$4 - \square$
yes (no)	**b)**	$\square + 12 >$
(yes) no	**c)**	$8 + \square > 10$
(yes) no	**d)**	$2 + \square > 3 \times 4$
(yes) no	**e)**	$9 > \square$
(yes) no	**f)**	$\square > 9$
(yes) no	**g)**	$p + 12 > 16$

Analysis of Question 10   Students should realize that 10c through 10g are inequalities but are true only for certain replacement values of the unknown. Those who answer **yes** to 10a should be reminded that 10a is an example of a mathematical expression, while 10b is not a well-formed inequality because it has no right side.

Guiding Class Discussion   The practice sheets and suggested activities in the Follow-up Instruction section under the concept **Understand the concept of inequality** may be helpful for those students having difficulty or needing more practice.

Focus of Question 11   Question 11 parallels question 6 but addresses the decomposition idea in the context of inequality.

Question 11

11.

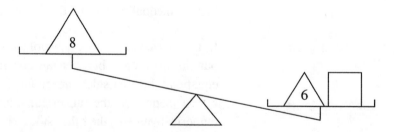

In the figure above, the weights on the right pan of the scale are heavier than the weight on the left pan. The □ in this figure stands for an unknown weight. Consider each of the following. Could the number sentence describe the figure?

yes (no)	**a)**	$8 < 6 + 1$
yes (no)	**b)**	$8 < 6 + 2$

(yes)	no	**c)**	$8 < 6 + 3$
yes	(no)	**d)**	$8 > 6 + 1$
yes	(no)	**e)**	$8 > 6 + 2$
yes	(no)	**f)**	$8 > 6 + 3$
(yes)	no	**g)**	$8 < 6 + \square$ when 4 is placed in the $\square$.
yes	(no)	**h)**	$8 > 6 + \square$ when 4 is placed in the $\square$.
yes	(no)	**i)**	$6 + 1 < 8$
(yes)	no	**j)**	$6 + 3 > 8$

Analysis of Question 11

For those students who have difficulty with one or more parts of question 11, the teacher might use the double-pan scale to model each inequality. In order to answer each part of this question, the students must determine what numbers could go in the $\square$ in the figure so that $8 < 6 + \square$. A **yes** answer to 11a may mean that the meaning of $<$ is not clear, while a **yes** answer to 11b or 11e requires the teacher to explain that the two sides are equal in this case. It should be emphasized that, although 11d is a true inequality, it does not describe the figure. Some students may have trouble seeing this, since the left pan in the figure is higher than the right pan. There may be an incorrect association between $>$ and the height of the left pan. Examining the students' responses to 11g and 11h will give the teacher further information about this misconception. Also, the students should be questioned to determine whether they realize that in 11g there are many numbers that can be placed in the $\square$ to make the inequality true. This fundamental difference between an inequality and an equation should be pointed out to the students. In 11i and 11j, the teacher should point out that it is possible to write an inequality in which the left side corresponds to the right pan of the scale. However, $>$ or $<$ must be adjusted accordingly. The inequality in 11i is incorrect since the right pan is heavier than the left, necessitating the use of $>$ instead of $<$ and the unknown weight must be greater than 2.

Guiding Class Discussion

The practice sheets and suggested activities in the Follow-up Instruction section under the concept **Understand the concept of inequality** may be helpful for those students having difficulty or needing more practice.

Focus of Questions 12–13

Question 12 continues the assessment of the concept of inequality at a higher level, since there are several terms on each side.

Question 13 involves the idea that a $\square$ (or literal term) in an inequality may have many values that will make that inequality true.

Questions 12–13

Questions 12–13 refer to the following information.

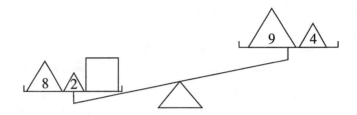

In the figure above, the weights on the left pan of the scale are heavier than the weights on the right pan. The □ in this figure stands for an unknown weight.

12. Write a number sentence that uses □ and > to describe the figure.

Answer: _8 + 2 + □ > 9 + 4_

13. Consider each of the following. Could the number be the weight of the □ in the figure?

yes	⬭no⬭	**a)**	1
yes	⬭no⬭	**b)**	3
⬭yes⬭	no	**c)**	5
⬭yes⬭	no	**d)**	7

Analysis of Questions 12–13

The teacher should question those students who write $10 + □ > 13$ as their number sentence in question 12 to be certain that they understand the inequality presented in the situation in the figure.

The teacher should discuss with the students the fact that 13b results in an equality, which is a different (but associated) number sentence.

Guiding Class Discussion

The practice sheets and suggested activities in the Follow-up Instruction section under the concept **Understand the concept of inequality** may be helpful for those students having difficulty or needing more practice.

Focus of Question 14

Question 14 assesses whether the students can recognize that the associated decomposition form of a true equation is also true (as in 14a, 14d, and 14e). The students should not perform any calculations to answer each part of the question since the goal is to have them analyze each equation by inspection only.

Question 14

14. Let □ and △ stand for two different numbers. Answer **yes** or **no** for each of the following questions. (Do *not* compute to determine your answer.)

(yes)	no	**a)**	If $235 + 94 = \square$, is $\square = 235 + 94$?
yes	(no)	**b)**	If $735 + 215 = \triangle$, is $735 = 215 + \triangle$?
yes	(no)	**c)**	If $\square + 183 = 381$, is $\square + 381 = 183$?
(yes)	no	**d)**	If $\triangle + 129 = 538$, is $538 = \triangle + 129$?
(yes)	no	**e)**	If $210 + \square = 482$, is $482 = 210 + \square$?

Analysis of Question 14

The students should see that in 14a, 14d, and 14e, when the left and right sides are interchanged, the equality is preserved. The difference between 14a, 14d, and 14e is that the □ or △ is located in a different position in the first equation in each question.

In 14b + and = have been interchanged, and in 14c the positions of the numbers have been switched. However, in both cases, inspection reveals that the second equation in each question cannot be true if the first one is.

A related source of difficulty for students may be in realizing that the number that goes into the □ or △ remains the same within a question, but is not necessarily the same in different questions. The teacher should question the students to ascertain whether they understand that in 14a, for example, the □ in the first equation contains the same hidden number that the □ in the second equation contains.

Guiding Class Discussion

The practice sheets and suggested activities in the Follow-up Instruction section under the concept **Understand the role of the equal sign in a decomposition** may be helpful for those students having difficulty or needing more practice.

Focus of Question 15

In question 15 students must determine if the left and right sides of each number sentence are equivalent; in 15a, 15b, and 15c, there is more than one number on the right side. This question also assesses the students' ability to hold an expression in suspension; for example, students should be able to work with the expression 8 feet + 5 inches in that form, rather than combining quantities to get 101 inches.

Question 15

15. Consider each of the following. Is the equation true?

(yes)	no	**a)**	3 feet + 5 feet + 5 inches = 8 feet + 2 inches + 3 inches
(yes)	no	**b)**	3 feet + 5 feet + 5 inches = 8 feet + 5 inches
yes	(no)	**c)**	3 feet + 5 feet + 5 inches = 13 feet + inches
yes	(no)	**d)**	3 feet + 5 feet + 5 inches = 13

Analysis of Question 15

In algebra, students frequently have trouble seeing that the answer to a question can be an expression such as $4x + 3y$. They may attempt to combine terms, arriving at $7xy$ or some other incorrect expression. They cannot accept a lack of closure in such questions.

Students who answer **yes** to 15c or 15d are adding the numbers on the left side of the equation and ignoring the units of measure. They should be encouraged to determine the total number of feet and total number of inches on each side of each equation and compare the resulting quantities to determine if the sides are equivalent.

Guiding Class Discussion

The practice sheets and suggested activities in the Follow-up Instruction section under the concept **Understand the role of the equal sign in an equivalence** may be helpful for those students having difficulty or needing more practice.

Practice sheet 2, in particular, involves a continuation of the ideas presented in question 15. The focus is on combining like terms to determine if each equation is true.

Focus of Question 16

This question assesses whether the students understand the effect on the inequality sign when the left and right sides are interchanged. As in question 14, the emphasis should be on analyzing each statement by inspection and not on performing calculations.

Question 16

16. Let □ stand for a number. Answer **yes** or **no** for each of the following questions. (Do *not* compute to determine your answer.)

(yes)	no	**a)**	If $133 + 40 > □$, is $□ < 133 + 40$?
yes	(no)	**b)**	If $133 + 40 > □$, is $□ > 133 + 40$?
yes	(no)	**c)**	If $□ + 235 < 567$, is $□ + 567 < 235$?
(yes)	no	**d)**	If $□ + 235 < 567$, is $567 > □ + 235$?
(yes)	no	**e)**	If $479 + □ < 852 + 140$, is $852 + 140 > 479 + □$?

Analysis of Question 16

In 16a, 16d, and 16e, it should be pointed out to the students that, unlike equations, when two sides of an inequality are interchanged, the inequality sign must be reversed to maintain a correct statement. Students who answer **yes** to 16b need to have this idea emphasized. In 16c students should understand that interchanging numbers changes the inequality.

Guiding Class Discussion

The practice sheets and suggested activities in the Follow-up Instruction section under the concept **Understand the similarities and differences between equations and inequalities** may be helpful for those students having difficulty or needing more practice.

Focus of Question 17

Question 17 addresses the concept of equality as an equivalence. In each part of this question, the focus is on the right side of the equation, since the left side does not change. And, since the multiplication and addition operations occur from left to right on the left side of the equation, the issue of order of operations does not complicate the solution of each part of the question.

---

17. Consider each of the following. Is the equation true?

(yes) no	**a)**	$2 \times 3 + 15 = 21$
(yes) no	**b)**	$2 \times 3 + 15 = 4 \times 4 + 5$
yes (no)	**c)**	$2 \times 3 + 15 = 2 + 3 + 15$
(yes) no	**d)**	$2 \times 3 + 15 = 10 + 11$
yes (no)	**e)**	$2 \times 3 + 15 = 6 + 12$

Analysis of Question 17

Students should be able to accept as true equations those equations whose right sides contain more than one term (as in 17b and 17d).

Guiding Class Discussion

The teacher can use the double-pan scale to model the equations for those students who are having difficulty with this concept. Multiplication can be modeled as repeated addition.

The practice sheets and suggested activities in the Follow-up Instruction section under the concept **Understand the role of the equal sign in an equivalence** may be helpful for those students having difficulty or needing more practice.

Focus of Questions 18–20

Questions 18–20 assess knowledge of similarities and differences between equations and inequalities. Students should solve these equations and inequalities using number facts since the emphasis in this unit is not on algebraic methods of solution.

Questions 18–20

Questions 18–20 refer to the following information.

$$\{1, 2, 3, 4, 5, 6, 7, 8, 9, 10\}$$

The numbers in the set above are possible replacements for the $\square$ in each of the following number sentences.

18. Which number or numbers will make the sentence $\square + 8 = 13$ true?

Answer: _5_

19. Which number or numbers will make the sentence $\square + 8 < 13$ true?

Answer: _1, 2, 3, 4_

20. Which number or numbers will make the sentence $\square + 8 > 13$ true?

Answer: _6, 7, 8, 9, 10_

Guiding Class Discussion

One difference between the equation in question 18 and the inequalities in questions 19 and 20 is that, although the equation has one solution, the inequalities have several solutions. It is important that students realize that an inequality which contains a missing number may be true for *many* different values of the missing number.

Another important point that the teacher should present to the students is that by uniting the answers in questions 18, 19, and 20, the entire set of possible replacements for the $\square$ is obtained. This point can be illustrated on a number line as follows.

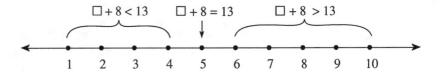

The double-pan scale could be used to model each inequality or equation with a replacement value from the set. For example, if 8 paper clips (or other objects each having the same weight) are placed on the left pan and 13 paper clips are placed on the right pan of the scale, the students should see that adding 5 paper clips to the left pan will cause the scale to balance (supporting the equation in question 18 as correct) while adding fewer than 5 paper clips or more than 5 paper clips will support the inequalities in questions 19 and 20, respectively, as correct.

The practice sheets and suggested activities in the Follow-up Instruction section under the concept **Understand the similarities and differences between equations and inequalities** may be helpful for those students having difficulty or needing more practice.

In particular, **practice sheet 3** provides two similar sets of equations and inequalities for the students to solve, choosing their solutions from a given set of replacement values.

Practice sheet 4 gives additional examples, organizing the solution sets in tabular form and emphasizing the fact that each element of the replacement set must satisfy either the equation or one of the two inequalities.

Focus of Questions 21–23

Questions 21–23 focus on the same concepts and ideas as questions 18–20. However, in this instance, the replacement set for the $\square$ has been divided into three subsets from which the students must choose the correct set for each sentence. Again, the students should observe that by uniting the three sets, the entire set of whole numbers displayed on the corresponding number line is obtained. The double-pan scale can be used to reinforce these concepts if needed.

Questions 21–23

Questions 21–23 refer to the following information.

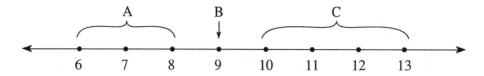

The points shown on the number line above can be represented by the following sets.

$$A \text{ is } \{6, 7, 8\}$$
$$B \text{ is } \{9\}$$
$$C \text{ is } \{10, 11, 12, 13\}$$

21. Which set contains a number or numbers that can be placed in the □ to make the number sentence $2 \times \square = 18$ true? Write the letter of the set in the space provided.

    Answer: *B*

22. Which set contains a number or numbers that can be placed in the □ to make the number sentence $2 \times \square < 18$ true?

    Answer: *A*

23. Which set contains a number or numbers that can be placed in the □ to make the number sentence $2 \times \square > 18$ true?

    Answer: *C*

Guiding Class Discussion

The teacher should have the students look at questions 18–23 and discuss with them the fact that there seems to be a relationship between the solution of the equation and the solutions of the two inequalities in each case. The students should investigate whether uniting separate solution sets will always result in the entire replacement set by working through practice sheets 3 and 4.

After the students have completed these activities, the teacher could discuss with them the nature of the relationship between an equation and its associated inequalities to be certain that they understand that the union of the solutions produces the entire replacement set.

The practice sheets and suggested activities in the Follow-up Instruction section under the concept **Understand the similarities and differences between equations and inequalities** may be helpful for those students having difficulty or needing more practice.

Focus of Questions 24–27   Questions 24–27 extend the relationship between equations and inequalities to those sentences that involve two different operations. As in previous questions involving two operations on the same side, multiplication occurs to the left of addition in the inequality.

Questions 24–27

Questions 24–27 refer to the following information.

$$\{1, 2, 3, 4, 5, 6, 7, 8, 9, 10\}$$

The numbers in the set above are possible replacements for the $\square$ in the inequality $2 \times \square + 1 > 9$.

24. What numbers in the set will make the inequality $2 \times \square + 1 > 9$ true?

Answer: _5, 6, 7, 8, 9, 10_

25. What numbers in the set will make the inequality $2 \times \square + 1 > 9$ false?

Answer: _1, 2, 3, 4_

26. What is the *largest* number in the set that will make the inequality $2 \times \square + 1 > 9$ false?

Answer: _4_

27. When you place your answer for question 26 in the $\square$ in the inequality $2 \times \square + 1 > 9$ and change the $>$ symbol to make a true sentence, what is the new sentence that will result?

Answer: _$2 \times \boxed{4} + 1 = 9$_

Guiding Class Discussion   The teacher should be aware of students who find it difficult to distinguish between the concepts of equality and inequality in question 26. It may help those students to work out the answer to this question on the double-pan scale to convince themselves that when 4 is placed in the $\square$, the left and right sides will each be 9 and the scale will balance. During the discussion, the teacher should point out that the idea of examining the related equation when solving an inequality is a good method to use to solve the inequality.

The teacher may wish to extend the discussion of this set of questions by having the students write the related inequality, $2 \times \square + 1 < 9$, and provide its solution.

The practice sheets and suggested activities in the Follow-up Instruction section under the concept **Understand the similarities and differences between equations and inequalities** may be helpful for those students having difficulty or needing more practice.

Focus of Questions 28–29   Questions 28 and 29 involve a common misuse of the equal sign.

Questions 28–29

28. Josh had 12 marbles. Luis gave him 8 more marbles. Then Josh gave Pam 5 marbles. How many marbles does Josh have now?

Consider each of the following. Is the statement a true description of the problem above?

(yes)	no	**a)**	$12 + 8 - 5 = 12 + 3 = 15$
yes	(no)	**b)**	$12 - 8 + 5 = 4 + 5 = 9$
yes	(no)	**c)**	$12 + 8 = 20 - 5 = 15$
yes	(no)	**d)**	$12 - 8 = 4 + 5 = 9$
(yes)	no	**e)**	$12 + 8 - 5 = 20 - 5 = 15$

29. Marta had 5 dollars. She earned 3 dollars delivering newspapers and received a 2-dollar allowance. How much does Marta have now?

Consider each of the following. Is the statement a true description of the problem above?

yes	(no)	**a)**	$5 + 3 = 5 + 3 + 2 = 10$
(yes)	no	**b)**	$5 + 3 + 2 = 8 + 2 = 10$
(yes)	no	**c)**	$5 + 3 + 2 = 5 + 5 = 10$
yes	(no)	**d)**	$5 - 3 - 2 = 2 - 2 = 0$
yes	(no)	**e)**	$5 - 3 = 2 - 2 = 0$

Analysis of Questions 28–29

Students who answer **yes** to 28c or 28d are using the equal sign to connect various steps of the problem, ignoring the equations involved. They assume that the first number to the right of the equal sign is the result of the calculations shown to the left of that equal sign and that they can continue to operate on that result with another calculation. This is also the case in 29a and 29e. These students should be reminded that in 28d, for example, $12 - 8 \neq 9$. It should also be pointed out to the students that 28a, 28e, 29b, and 29c are true sentences because the expressions on each side of the equal sign are equal. In 29d, the sentence is true but the answer is **no** because it is not a true description of the problem.

Guiding Class Discussion

The practice sheets and suggested activities in the Follow-up Instruction section under the concept **Understand the role of the equal sign in an equivalence** may be helpful for those students having difficulty or needing more practice.

Focus of Questions 30–31

Question 30 assesses the students' ability to write an inequality for a real-world situation, and question 31 assesses understanding of the concept of inequality.

Questions 30–31

Questions 30–31 refer to the following information.

Larry and Mike are planning a party at Vito's Pizza Parlor. Vito will give a free pizza on the condition that the total group (including Larry and Mike) has more than 10 people in it.

30. If □ stands for the number of friends that Larry and Mike will invite to the party, what is an inequality that expresses the condition that the total group (including Larry and Mike) has more than 10 people in it?

    Answer: $\square + 2 > 10$

31. Larry and Mike are thinking about inviting 7, 8, 9, 10, 11, or 12 friends to their party. Which of these numbers is the *smallest* number of friends they could invite and still get a free pizza?

    Answer: 9

Guiding Class Discussion

The teacher should discuss with the students that by looking at the equation □ + 2 = 10, one could determine that 9 is the smallest number that makes the inequality □ + 2 > 10 true.

The discussion could then be extended to the inequality □ + 2 < 10, and the students could be asked to determine the greatest number that makes this inequality true.

The practice sheets and suggested activities in the Follow-up Instruction section under the concept **Construct numerical inequalities** may be helpful for those students having difficulty or needing more practice.

Practice sheet 5 contains several other word problems that generate inequalities, similar to questions 30 and 31, and may be used for additional practice if needed.

Follow-up Instruction

After the Instructional Assessment has been used to pinpoint students' conceptual weaknesses, the teacher may wish to use the following suggested activities and practice sheets to help correct those misconceptions.

For several of the activities in the unit, it will be useful for each student or pair of students in a class to have a set of "tiles." These tiles can be cut from heavy-gauge paper. Each tile should be a square whose side is at least $1\frac{1}{2}$ inches in length and should contain a whole number (from 1 through 50, for example) or one of the symbols $+$, $-$, $\times$, $\div$, $=$, $<$, or $>$. Each set of tiles should contain several copies of the same tile so that equations or inequalities which contain identical numbers or symbols may be formed. For example, the teacher may want to include in each set three copies of each of the fifty number tiles and five copies of each of the seven symbol tiles.

Understand the concept of equality

Activity. An activity that relates to question 1 is to have the students use their sets of tiles to form true equations. The teacher could direct the students to form equations with one operation on the left side and one number on the right, with one number on the left side and one operation on the right, or with a different operation on each side. Note that two important ideas are involved—forming an equation and checking that the equation is true. A double-pan scale or other balance device and a set of weights might be used to confirm or refute various conjectures regarding equations that contain the operation of addition, for example.

Practice sheets 1 and 2, discussed under the concepts Understand the role of the equal sign in a decomposition and Understand the role of the equal sign in an equivalence, respectively, address this concept, as do questions 1–4 in the Follow-up Assessment.

Understand the role of the equal sign in a decomposition

Activity. The teacher could spend additional time developing the concept of decomposition by having each student in the class make up a true equation and write it on a sheet of paper. The papers could then be exchanged with other students who would write the equation in different ways. Many alternate forms of an equation could be generated by using words and phrases that mean the same as =, by writing the associated decomposition form of the equation, and then by using the words and phrases in the decomposition form. The teacher may also want to explore with the class other words and phrases that mean the same as the equal sign. (Your textbook and other mathematics textbooks could be used as references.)

Activity. The students can use some of the tiles to reinforce the concept of decomposition. The teacher should direct the students to use either a + or × tile, an = tile, and three number tiles to form a correct equation. Then they could form as many other correct equations as possible. For example, if the student forms the equation $6 \times 7 = 42$, the other equations that should be generated are $7 \times 6 = 42$, $42 = 6 \times 7$, and $42 = 7 \times 6$. (Note: If only a limited number of tiles are available, the students can write each equation on a sheet of paper as it is formed and immediately reuse the tiles.) After all possible equations for one set of five tiles are generated, the process can be repeated using different number tiles.

For a follow-up activity, the teacher could have the students examine all equations which were formed in the activity above and pair the equations in which the left side of the first equation in the pair is identical to the right side of the second equation and the right side of the first equation is identical to the left side of the second equation. The idea is to help the students realize that equations such as $6 \times 7 = 42$ and $42 = 6 \times 7$ are the same and that $7 \times 6 = 42$ and $42 = 7 \times 6$ are the same; *i.e.*, the equality is maintained when the left and right sides of an equation are interchanged.

Activity. After the students have used numerical equations to gain an understanding of decomposition, the activities above could be extended by having the students "hide" a number in the equation they have formed. For example, if one turns over the 7 tile in the equation $6 \times 7 = 42$, the equation $6 \times \square = 42$ is obtained. This corresponds to $6 \times \square = 42$, from which the equation $42 = 6 \times \square$ can be formed. This extension activity focuses on decomposition equations that contain an unknown; however, since the students already know what goes into the box because they turned over the tile, they may have a better understanding of the role of the $\square$ in the equation and the concept of equality in $6 \times \square = 42$ and $42 = 6 \times \square$. The teacher should point out that $\square \times 6 = 42$, $42 = \square \times 6$, $\square \times 7 = 42$, $7 \times \square = 42$, $42 = \square \times 7$, $42 = 7 \times \square$, $7 \times 6 = \square$, $6 \times 7 = \square$, $\square = 7 \times 6$, and $\square = 6 \times 7$ can also be formed from the same tiles.

The teacher may want to provide some additional instruction in the concept of decomposition as related to equality by having the students use their sets of tiles to form true numerical equations and then hide one number in each equation by turning over a tile. After writing the resulting equation on a sheet of paper, the students could then write the equation that is formed when the left and right sides have been interchanged. To focus this activity on decomposition, the teacher should be certain that the right side of each initial equation contains one number or a □ only.

Practice sheet 1 gives a more complicated decomposition activity in which students complete a table of values for an equation in two variables. In this type of exercise, when two operations such as multiplication and addition are used, the multiplication occurs to the left of the addition in the equation, since order of operations may otherwise complicate the exercise (*i.e.*, students usually read from left to right and perform the operations one at a time and in that order).

Questions 5–9 in the Follow-up Assessment also deal with this concept.

Understand the role of the equal sign in an equivalence

Activity. The double-pan scale can be used to model equations that have two or more terms on each side of the equal sign. Given a certain distribution of weights on a balanced scale, the students can be asked to write an equation that correctly describes the situation (as in questions 7 and 8) and students can also check that the equations they have written without the aid of a scale are correct by modeling the equations on the scale.

The teacher could also use the scale to model the idea of an unknown in an equation. The students could be presented with a balanced scale in which one of the weights has been concealed. After the students have written an equation in terms of an unknown, the □, they could be asked to determine what number should be placed in the □. Their responses could then be checked by revealing the concealed weight.

On **practice sheet 2** students must group like terms on one side of the equal sign to check the truth of given equations. The use of labels other than units of measure may help some students understand what is meant by "like terms." They may then find it easier to simplify such expressions as $3x + 5x + 4y + 6y$.

Follow-up Assessment questions 10–13 also relate to this concept.

Understand the concept of inequality

Activity. To develop an understanding of the concept of inequality by examining how inequalities and equations are related, the students should first model an inequality such as $8 < 5 + 10$ on the double-pan scale by putting 8 identical objects on the left pan and 5 objects + 10 objects on the right pan. (The model should be set up so that the left pan of the scale corresponds to the left side of the inequality or equation. This will be less confusing for the students.) Next, they should be asked to determine how many weights must be added to the lighter side to make the scale balance. Finally, they should be asked to write the equation that describes the balanced scale. In this case, the resulting equation could be $8 + 7 = 5 + 10$. This activity helps the students to analyze an inequality in terms of the associated equation.

Activity. Another activity that can be used to help students gain a better understanding of the concept of inequality makes use of the sets of tiles that were made earlier. The tiles can be used by the students to form true inequalities which contain numbers only, and then an inequality containing a variable can be formed by hiding one number in the numerical inequality. The students could then determine for which whole numbers the inequality will be true.

Activity. The double-pan scale can be also used to model inequalities. In this activity the students should be able to see easily that the greater side of the inequality corresponds to the heavier side of the scale.

As a follow-up to the ideas presented in question 11, the students could be given several inequalities and asked to draw a double-pan scale that models each inequality. This would give the students additional practice in realizing that the lesser side of an inequality corresponds to the pan that is vertically higher in the figure.

The double-pan scale can also be used by the students, along with paper clips or marbles, for example, to model various inequalities that have several numbers on both sides. The students could be presented with an unbalanced scale that has two different groups of marbles on each pan and asked to write an inequality as in question 12; or the students could write their own inequalities with two or more numbers on each side and model them on the double-pan scale.

Practice sheets 3–5 and questions 14–18 in the Follow-up Assessment also address this concept.

Understand the similarities and differences between equations and inequalities

Practice sheets 3 and 4 emphasize that for a given equation and its associated inequalities, uniting the separate solution sets will result in the entire replacement set.

Follow-up Assessment questions 19–28 deal with this concept.

Construct numerical inequalities

Practice sheet 5 presents real-world situations that can be represented by inequalities. Students are also asked to determine the greatest member of the replacement set that satisfies a given condition.

Follow-up Assessment questions 29 and 30 cover this concept.

Bibliography of Resource Materials

Herscovics, Nicolas, and Carolyn Kieran. "Constructing Meaning for the Concept of Equation." *Mathematics Teacher 73* (November 1980): 572–580.

Kieran, Carolyn. "Concepts Associated with the Equality Symbol." *Educational Studies in Mathematics 12* (August 1981): 317–326.

Answers

Answers to Instructional Assessment

1. a) no
b) no
c) yes
d) no
e) no
f) yes
g) yes
h) yes

2. a) yes
b) no
c) no
d) no
e) yes
f) yes
g) no
h) yes
i) yes
j) no
k) yes

3. a) yes

b) yes
c) yes
d) no
e) yes
f) yes
g) no
h) no
i) yes

4. $20 + 16 = 36$

5. Answers will vary. For example, $36 = 20 + 16$

6. a) yes
b) yes
c) no
d) no

7. Answers will vary.

For example, $4 + 7 + \square = 10 + 6$

8. 5

9. a) no
b) no
c) yes
d) yes
e) no
f) yes

10. a) no
b) no
c) yes
d) yes
e) yes
f) yes
g) yes

11. a) no
b) no

c) yes
d) no
e) no
f) no
g) yes
h) no
i) no
j) yes

12. Answers will vary. For example, $8 + 2 + \square > 9 + 4$

13. a) no
b) no
c) yes
d) yes

14. a) yes
b) no

c) no
d) yes
e) yes

15. a) yes
b) yes
c) no
d) no

16. a) yes
b) no
c) no
d) yes
e) yes

17. a) yes
b) yes
c) no
d) yes
e) no

18. 5

19. 1, 2, 3, 4
20. 6, 7, 8, 9, 10

21. *B*
22. *A*
23. *C*
24. 5, 6, 7 8, 9, 10
25. 1, 2, 3, 4
26. 4
27. $2 \times 4 + 1 = 9$
28. a) yes
b) no
c) no
d) no
e) yes
29. a) no
b) yes
c) yes
d) no
e) no
30. $\square + 2 > 10$
31. 9

Answers to Practice Sheets

Practice sheet 1

Answers will vary.

1. For example, some possible
   number pairs for □ and △ are

□	△
1	34
2	32
3	30
4	28
5	26

2. For example, some possible
   number pairs for □ and △ are

□	△
10	12
9	9
8	6
7	3
6	0

Practice sheet 2

   1. yes
   2. no
   3. no
   4. no
   5. yes
   6. no
   7. yes
   8. no
   9. yes
   10. no
   11. no
   12. no

Practice sheet 3

A. {7}
B. {8, 9, 10}
C. {1, 2, 3, 4, 5, 6}

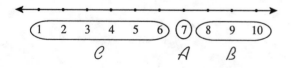

D. {4}
E. {1, 2, 3}
F. {5, 6, 7, 8, 9, 10}

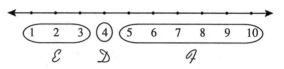

Practice sheet 4

2. b) $5 \times \square > 35$; {8, 9, 10}
   c) $5 \times \square < 35$; {1, 2, 3, 4, 5, 6}
   d) yes
3. a) $10 - \square = 4$; {6}
   c) $10 - \square < 4$; {7, 8, 9, 10}
   d) yes
4. a) $\square + 9 = 14$; {5}
   b) $\square + 9 > 14$; {6, 7, 8, 9, 10}
   d) yes
5. b) $2 \times \square + 3 > 9$; {4, 5, 6, 7, 8, 9, 10}
   c) $2 \times \square + 3 < 9$; {1, 2}
   d) yes
6. a) $\square + 7 - 2 = 10$; {5}
   c) $\square + 7 - 2 < 10$; {1, 2, 3, 4}
   d) yes
7. a) $30 \div \square = 6$; {5}
   b) $30 \div \square > 6$; {1, 2, 3, 4}
   d) yes

Practice sheet 5

1. For example, $\square + 6 < 18$.
2. 11
3. For example, $\triangle + 12 < 25$.
4. 12
5. $15 + \triangle > 20 + \square$
6. 6

Answers to Follow-up Assessment

1. a) yes
b) no
c) no
d) no
e) yes
f) yes
g) no
h) yes
2. a) yes
b) no
c) yes
d) no
e) no
f) yes
g) no
h) yes
i) no
j) no
k) yes

3. a) no
b) yes
c) no
d) no
e) yes
4. a) no
b) no
c) no
d) yes
e) no
5. a) yes
b) no
c) no
d) yes
e) yes
f) no
g) yes
h) yes
6. $9 \times 5 = 45$

7. For example, $45 = 9 \times 5$
8. a) yes
b) yes
c) no
d) no
e) no
9. a) no
b) yes
c) yes
d) yes
e) no
10. For example, $8 + 12 = \square + 9 + 2$
11. 9
12. a) yes

b) yes
c) no
d) no
13. a) yes
b) yes
c) no
d) yes
e) no
14. a) no
b) yes
c) no
d) yes
e) no
f) yes
15. a) yes
b) yes
c) no
d) yes
e) yes

f) no
g) no
16. a) no
b) no
c) yes
d) no
e) no
f) no
g) no
h) no
i) no
j) yes
17. For example, $11 + 5 < \square + 7 + 2$
18. a) no
b) no
c) no

d) yes
19. a) yes
b) no
c) yes
d) yes
e) yes
20. 7
21. 8, 9, 10
22. 1, 2, 3, 4, 5, 6
23. B
25. C
25. A
26. 1, 2, 3
27. 4
28. $4 \times \boxed{4} - 3 = 13$
29. $\square + 3 < 12$
30. 8

ALGE**BRIDGE**™

Operations on

Equations and

Inequalities

Operations on Equations and Inequalities

Introduction

The operations used to solve equations and inequalities can be divided into two groups: (1) those related to simplifying one member (side) of the equation or inequality and (2) those related to performing the same operation on both members of the equation or inequality. Since the motivation for performing operations on both sides of an equation is to find a solution to the equation, this unit views equations as equivalences where both sides are numerically equal for some value of the unknown.

In the context of simple linear equations and inequalities that involve whole numbers, this unit develops the student's ability to simplify a member of an equation or an inequality by the following processes:

- group or combine like terms;

- apply the rules that govern the order of operations;

- apply the properties of commutativity, associativity, and distributivity.

This unit also builds the student's ability to transform an equation or an inequality to an equivalent form by the following:

- select appropriate inverse operations;

- perform the same operation on both sides.

This unit does not presume a knowledge of negative numbers.

Cross-References to
Pre-Algebra and Algebra
Textbook Topics

What follows is a compilation of chapter titles and index entries that are found in typical pre-algebra and algebra textbooks. The concepts and abilities addressed in this unit could be used to enhance the instruction on these topics.

algebraic expressions	**open sentences**
basic operations	**order of operations**
formulas	**solving equations**
integers and equations	**solving inequalities**
number properties and equations	**transforming equations**

List of Terms

The following is a list of words and phrases that are used frequently in discussing the topics covered in this unit.

The teacher may wish to add other appropriate terms.

associative	**like (similar) terms**
balanced	**order of operations**
commutative	**side (member) of an equation**
conversion	**simplify**
distributive	**solve (an equation or an inequality)**
equal	**solution**
equation	**substitution**
equivalent	**term**
evaluate	**undo an operation**
expression	**(inverse operation)**
inequality	**variable**

Classification by Concept

Target Concepts and Abilities	Questions in Instructional Assessment	Questions in Follow-up Assessment	Practice Sheets
Group or combine like terms	1, 2, 3, 4, 12, 13, 14, 28	1, 2, 9, 23	1, 6
Apply the rules that govern the order of operations	5, 6, 7, 8, 9, 10, 11, 28	3, 4, 5, 6, 7, 8, 23	2, 3, 4, 5
Apply the properties of commutativity, associativity, and distributivity	4, 5, 7, 8, 9, 10, 11, 15, 27	2, 3, 5, 6, 7, 8, 10, 22	2, 3, 4, 5, 6, 8
Select appropriate inverse operations	15, 22, 23, 24, 25, 26, 28	10, 17, 18, 19, 20, 21, 23	8
Perform the same operations on both sides of an equation or inequality	16, 17, 18, 19, 20, 21, 25, 26, 27, 28	11, 12, 13, 14, 15, 16, 20, 21, 22, 23	7, 8

Classification by Question

Target Concepts and Abilities		Group or combine like terms	Apply the rules that govern the order of operations	Apply the properties of commutativity, associativity, and distributivity	Select appropriate inverse operations	Perform the same operations on both sides of an equation or inequality
Question Number or Practice Sheet Number	1	IA FA PS				
	2	IA FA	PS	FA PS		
	3	IA	FA PS	FA PS		
	4	IA	FA PS	IA PS		
	5		IA FA PS	IA FA PS		
	6	PS	IA FA	FA PS		
	7		IA FA	IA FA		PS
	8		IA FA	IA FA PS	PS	PS
	9	FA	IA	IA		
	10		IA	IA FA	FA	
	11		IA	IA		FA
	12	IA				FA
	13	IA				FA
	14	IA				FA
	15			IA	IA	FA
	16					IA FA
	17				FA	IA
	18				FA	IA
	19				FA	IA

IA - Questions in Instructional Assessment
FA - Questions in Follow-up Assessment
PS - Practice Sheets

257

Classification by Question

Target Concepts and Abilities	Group or combine like terms	Apply the rules that govern the order of operations	Apply the properties of commutativity, associativity, and distributivity	Select appropriate inverse operations	Perform the same operations on both sides of an equation or inequality
Question Number or Practice Sheet Number 20				FA	IA FA
21				FA	IA FA
22			FA	IA	FA
23	FA	FA		IA FA	FA
24				IA	
25				IA	IA
26				IA	IA
27			IA		IA
28	IA	IA		IA	IA

IA - Questions in Instructional Assessment
FA - Questions in Follow-up Assessment
PS - Practice Sheets

Instructional Assessment

The questions in both the Instructional and Follow-up Assessment Instruments for this unit have different formats appropriate to the nature of the concepts addressed in the questions.

It is necessary to introduce the students to these formats since they may have limited experience with different types of responses.

If there is not sufficient time to review examples of the different question types, review the directions below with the students before they work the *Algebridge*™ exercises.

Directions

There are three types of questions in this exercise.

One type asks you to choose the answer to the problem and mark the corresponding letter, A, B, C, D, or E, on your answer sheet.

The second type gives you several choices related to a specific question. You must answer **yes** or **no** for each of these choices.

The third type asks you to write your own answer to the problem.

Focus of Questions 1–4

Foreshadowing expressions such as $5x + 3y + 2y + 7x$ in algebra, questions 1–4 address the problems students have in combining like terms. The first question contains concrete items, while the others are more abstract. Question 2 assesses whether a student views a constant term and a term with a variable as unlike terms, while question 3 assesses whether students distinguish between terms with different variables. Question 4 assesses the abilities addressed in questions 2 and 3, but with a more complex expression.

Question 1

1. Consider each of the following. Is the expression in column A equal to the expression in column B?

			Column A	Column B
(yes)	no	a)	5 lbs + 3 oz + 2 oz + 7 lbs	12 lbs + 5 oz
(yes)	no	b)	$7 + 20¢ + $4 + 50¢	$11 + 70¢
yes	(no)	c)	3 ft + 2 in + 5 yds + 3 in	8 ft + 5 in
yes	(no)	d)	4 m + 6 cm − 2 cm	8 m

Analysis of Question 1

Students who answer 1c incorrectly should be reminded to pay attention to the units of measure. Students who answer 1d incorrectly may be ignoring the units of measure altogether.

Guiding Class Discussion

Notice that conversions are not necessary in these examples because the expressions in Column B are not converted. However, a discussion about conversion might be very useful. Students may realize they can combine pounds and ounces by converting to ounces, and feet and inches by converting to inches, etc. They will probably want to combine the terms in part 1b to $11.70. This is a good place to stress two important points:

(1) The reason they can combine the terms is because they know the conversion.

(2) The conversion must be done before combining terms.

For example, 8 ft can be combined with 5 in if 8 ft is first converted to 96 in. The reason they can be combined is that 5 in and 96 in are "like terms."

It is possible that the abbreviations used for the units of measure may cause some confusion for some students. It would be helpful if the teacher rewrote the expression with "pounds," "feet," and so on for those students.

The practice sheets and suggested activities in the Follow-up Instruction section under the concept **Group or combine like terms** may be helpful for those students having difficulty or needing more practice.

Questions 2–3

2. Consider each of the following expressions. If □ stands for a number, is the expression equal to 5 · □ + 4 · □ + 6?

(yes)	no	a)	9 · □ + 6
yes	(no)	b)	5 · □ + 10 · □
yes	(no)	c)	15 · □

3. Consider each of the following expressions. If □ and △ each stand for a different number, is the expression equal to $7 \cdot \square + 8 \cdot \triangle + 2 \cdot \square$?

yes	(no)	**a)**	$17 \cdot \triangle \cdot \square$
(yes)	no	**b)**	$9 \cdot \square + 8 \cdot \triangle$

Analysis of Questions 2–3

Questions 2 and 3 include various incorrect methods of combining unlike terms that students sometimes use. A **yes** answer for 2b may mean the students feel they have to "do something" with the 6 by combining it with $4 \cdot \square$. A student who answers 2c or 3a incorrectly may believe there has to be only one term in the answer, as is often true in arithmetic problems.

Guiding Class Discussion

Students should be reminded that △ and □ may represent different unknown numbers so they are not like terms. A conversion is not possible as was the case in question 1. Therefore, the terms involving □ and △ cannot be combined with each other or with numerals.

It would be worthwhile at this point to discuss the distributive property. Students who know, for example, that $(4 + 5)3 = 4 \cdot 3 + 5 \cdot 3$ may be convinced by that argument that $8 \cdot \square + 9 \cdot \square = (8 + 9) \cdot \square = 17 \cdot \square$, but that the distributive property does not apply in the case of $8 \cdot \square + 6 \cdot \triangle$.

Other expressions the teacher might want to use as sample exercises include ones such as $11 \cdot \square + 2 + 3 \cdot \square + 4$ and $9 + 3 \cdot \square + 6 \cdot \triangle + 5 \cdot \triangle + 1$.

The practice sheets and suggested activities in the Follow-up Instruction section under the concept **Group or combine like terms** may be helpful for those students having difficulty or needing more practice.

Question 4

4. Consider each of the following expressions. If □ and △ each stand for a different number, is the expression equal to $8 \cdot \square + 6 \cdot \triangle + 9 \cdot \square + 3 + 7$?

(yes)	no	**a)**	$17 \cdot \square + 6 \cdot \triangle + 10$
yes	(no)	**b)**	$33 \cdot \square \cdot \triangle$
yes	(no)	**c)**	$23 \cdot \square \cdot \triangle + 10$
yes	(no)	**d)**	$27 \cdot \square + 6 \cdot \triangle$
yes	(no)	**e)**	$17 \cdot \square + 16 \cdot \triangle$
(yes)	no	**f)**	$17 \cdot \square + 6 \cdot \triangle + 3 + 7$
yes	(no)	**g)**	$20 \cdot \square + 13 \cdot \triangle$

Analysis of Question 4

In 4b all of the numbers are added and the □ and △ are combined, as in 2c and 3a. 4c is similar to 3a in that the students are not distinguishing

261

between the two different placeholders if they answer **yes**. A **yes** answer for either 4d or 4e means the student is adding the 10 to either the $17 \cdot \square$ or the $6 \cdot \triangle$. A **yes** answer for 4g may mean the students cannot decide what to do with the 3 and the 7, so they combine the 3 with $17 \cdot \square$ and the 7 with $6 \cdot \triangle$.

Guiding Class Discussion

The practice sheets and suggested activities in the Follow-up Instruction section under the concepts **Group or combine like terms** and **Apply the properties of commutativity, associativity, and distributivity** may be helpful for those students having difficulty or needing more practice.

In particular, on **practice sheet 1** the students use the technique of substitution to demonstrate which of the expressions in question 4 are combined incorrectly.

Focus of Questions 5–7

When simplifying the members of equations or inequalities, students often have difficulty with the order of operations. Questions 5–7 focus on the order of operations and assess the students' ability to combine several numerical terms. Question 5 focuses on addition and subtraction and questions 6 and 7 on addition and multiplication.

Question 5

5. Suppose $\square$ stands for the number of marbles that John has at the beginning of the day. During the day he wins 4 more marbles and then loses 3 of them on the way home because he has a hole in his pocket.

The number of marbles John then has can be represented by the expression $\square + 4 - 3$.

Study each of the following. Is the expression equal to $\square + 4 - 3$?

(yes)	no	a)	$\square - 3 + 4$
yes	(no)	b)	$\square + 3 - 4$
(yes)	no	c)	$4 + \square - 3$
yes	(no)	d)	$3 + 4 - \square$
yes	(no)	e)	$3 - 4 + \square$

Analysis of Question 5

In reading the expressions, the terms "add 4" and "subtract 3" should be stressed. By relating these expressions to a concrete setting, students can be led to realize that it does not matter which of these operations is done first and that the expression in 5a is equal to the given expression. Students who answer **no** to 5a may be focusing on the sequence of events, rather than on the final result of the operations. The teacher can point out that if the sequence of events were different (for example, losing 3 marbles followed by winning 4 marbles), the resulting number of marbles would be the same. Therefore, the value of the expression in 5a does represent the final number of marbles. By using the terms "add 4" and "subtract 3," the teacher can avoid the necessity of discussing negative numbers, especially if the students are not comfortable with them.

If students answer 5b incorrectly, they may not understand that subtraction is not commutative. The teacher may want to discuss the commutative property at this time.

If students answer **yes** to 5d, they may be arbitrarily exchanging terms without regard to the operations. By relating this back to the concrete setting, the teacher can point out why □ is not a quantity that should be subtracted. Those students who answer 5e incorrectly may think that the meaning of an expression does not change when it is read from right to left.

Guiding Class Discussion

To discover if the source of student difficulties with the question is the abstract symbol □, the teacher can present situations that can be represented by expressions using numbers only. For example, John has 7 marbles at the beginning of the day. He loses 5 marbles to Beth and wins 6 marbles from Paul.

The practice sheets and suggested activities in the Follow-up Instruction section under the concepts **Apply the rules that govern the order of operations** and **Apply the properties of commutativity, associativity, and distributivity** may be helpful for those students having difficulty or needing more practice.

In particular, on **practice sheet 2**, question 1, students use the technique of substitution to check the validity of the alternate expressions given in this assessment question. Questions 2 and 3 on practice sheet 2 give similar word problems emphasizing combining terms.

Question 6

6. What is the value of the expression $5 + 2 \times 3$?

Answer: *11*

Analysis of Question 6

Students who answer 21 to question 6 are performing the operations in order from left to right. They should be reminded of the rule that multiplication and division should be done before addition and subtraction.

Guiding Class Discussion

The practice sheets and suggested activities in the Follow-up Instruction section under the concept **Apply the rules that govern the order of operations** may be helpful for those students having difficulty or needing more practice.

For a related activity, the teacher may wish to use **practice sheet 3**, which contrasts how a 4-function calculator and a scientific calculator each handle the question of the order of operations.

Question 7

7. Suppose □ stands for the number of students sitting in each of 6 rows in a classroom. There are 3 students standing because they have no seats.

The total number of students in the classroom can be represented by the expression $\square \times 6 + 3$.

Consider each of the following. Is the expression equal to $\square \times 6 + 3$?

yes (no)	**a)**	$\square \times 9$
yes (no)	**b)**	$\square + 3 \times 6$
yes (no)	**c)**	$\square \times 3 + 6$
(yes) no	**d)**	$6 \times \square + 3$
(yes) no	**e)**	$3 + 6 \times \square$

Analysis of Question 7

In discussing question 7, the teacher should use the terms "multiplying by 6" and "adding 3." Then it is easier for students to see that 7a and 7c are not the same as the given expression.

Students who answer 7b and 7c incorrectly may be confused about commutativity. In 7b it should be pointed out which quantity should be multiplied by 6. Referring back to the concrete setting can help make this more plausible to the students. In 7c students may think they have correctly applied the commutative property of addition. If students answer **no** to 7e, they may be performing the operations sequentially from left to right. Those who answer both question 6 and question 7e incorrectly are probably doing just that.

Guiding Class Discussion

The practice sheets and suggested activities in the Follow-up Instruction section under the concepts **Apply the rules that govern the order of operations** and **Apply the properties of commutativity, associativity, and distributivity** may be helpful for those students having difficulty or needing more practice.

In particular, on **practice sheet 4**, question 1 uses the technique of substitution to check for equivalence of the alternate expressions (7a-e) with the expression given in question 7. Questions 2 and 3 on the practice sheet give similar word problems involving multiplication and addition.

Focus of Questions 8–10

Questions 8–10 assess whether students know they can do a series of additions and subtractions in several different orders and get the same results. The operations in questions 8 and 9 are presented in the context of an equation to assess whether students understand the principle in this setting. Question 9 differs from question 8 in the sequence of operations considered. Question 10 assesses the students' understanding of the same concept, but in the context of an inequality. The teacher should emphasize that the same rules for the order of operations that apply to equations also apply to inequalities.

Questions 8–9

8. Suppose $\square + 10 + 7 - 4 - 5 = 11$ is true when $\square$ represents some number.

    In each of the following some of the numbers on the left side of the equation above have been combined in some way.

    Consider each equation. Were the numbers combined correctly?

(yes)	no	**a)**	$\square + 17 - 4 - 5 = 11$
(yes)	no	**b)**	$\square + 13 - 5 = 11$
(yes)	no	**c)**	$\square + 17 - 9 = 11$
(yes)	no	**d)**	$\square + 8 = 11$
(yes)	no	**e)**	$\square + 10 + 3 - 5 = 11$

9. Suppose $\square - 5 + 9 - 3 + 4 = 8$ is true when $\square$ represents some number.

    In each of the following, some of the numbers on the left side of the equation above have been combined in some way.

    Consider each equation. Were the numbers combined correctly?

(yes)	no	**a)**	$\square + 13 - 8 = 8$
(yes)	no	**b)**	$\square + 4 + 1 = 8$
(yes)	no	**c)**	$\square + 5 = 8$
yes	(no)	**d)**	$\square - 14 - 7 = 8$

Analysis of Question 9

If students understand that the equation given in question 9 involves "subtracting 5, adding 9, subtracting 3, and adding 4," they may see that they can do these operations in any order. For this reason, the teacher should stress this terminology and reinforce the notion that addition and subtraction have equal priority in the order of operations. In 9a, the additions are done first, then the subtractions. In 9b, the operations are done two at a time, and in 9c the calculation on the left side is complete.

Students who answer 9d incorrectly are giving priority to addition and doing it first by mentally inserting parentheses, perhaps without realizing it. The teacher can point out how parentheses would change this expression and explain that $\square - 5 + 9 - 3 + 4 \neq \square - (5 + 9) - (3 + 4)$.

Guiding Class Discussion

The practice sheets and suggested activities in the Follow-up Instruction section under the concepts **Apply the rules that govern the order of operations** and **Apply the properties of commutativity, associativity, and distributivity** may be helpful for those students having difficulty or needing more practice.

In particular, **practice sheet 5** gives similar examples of equations containing strings of additions and subtractions to be combined using the rules that govern the order of operations.

Question 10

10. Let □ stand for a number in the inequality $\square + 8 - 3 + 2 > 9$.

Consider each of the following inequalities. Could it be the result of combining numbers in the inequality above?

(Remember that "□ > △" means "□ is greater than △.")

yes	(no)	**a)**	$\square + 3 > 9$
(yes)	no	**b)**	$\square + 7 > 9$
yes	(no)	**c)**	$\square + 8 - 5 > 9$
(yes)	no	**d)**	$\square + 5 + 2 > 9$

Analysis of Question 10

Students may be inserting parentheses in the given inequality if they are answering **yes** to question 10a or 10c. For example, $\square + 8 - (3 + 2) > 9$. Students who answer **yes** to only one of these may not even realize they are inserting parentheses.

If students are using substitution of values as a method to answer parts of question 10, they may be getting incorrect results. Since many different values will make the original inequality true, they may be choosing one that also makes the other inequalities true. For example, the original inequality is true when □ is replaced by 10. So are 10a–10d! The teacher can use 4 for the □ so that only 10b and 10d are true. It is important to point out these inconsistencies to students who used this method and to warn them against such an approach. The fact that there will usually be many numbers that make an inequality true should be emphasized.

Guiding Class Discussion

The practice sheets and suggested activities in the Follow-up Instruction section under the concepts **Apply the rules that govern the order of operations** and **Apply the properties of commutativity, associativity, and distributivity** may be helpful for those students having difficulty or needing more practice.

Focus of Question 11

Question 11 assesses the students' ability to work with several of the concepts of the previous questions simultaneously. For example, can students solve an equation using the correct order of operations? Do they understand what to do when the □ is in different positions? Can they deal with operations on the right side of the equation?

Question 11

11. In each of the following, what number belongs in the $\square$ to make both sides of the equation equal to the same number?

a) $\square + 2 + 7 = 10 + 5$ Answer: _6_

b) $8 + 2 + \square = 18 - 6$ Answer: _2_

c) $10 = \square - 3 + 1$ Answer: _12_

d) $9 + 7 = 3 + \square - 2$ Answer: _15_

Analysis of Question 11

If students answer any part of question 11 incorrectly, the teacher should try to determine which of these concepts are causing difficulty.

Students should be able to solve these equations without using any formal equation-solving techniques. They can perform the indicated additions and subtractions and then substitute different values for $\square$ until they find the one that works.

Guiding Class Discussion

The teacher can make up more examples of the type in question 11 to allow the students to practice finding solutions to equations.

The practice sheets and suggested activities in the Follow-up Instruction section under the concepts **Apply the rules that govern the order of operations** and **Apply the properties of commutativity, associativity, and distributivity** may be helpful for those students having difficulty or needing more practice.

Focus of Questions 12–14

The purpose of questions 12 and 13 is to assess the students' understanding of the basic facts that $\square + \square = 2 \cdot \square$ and $\square - \square = 0$ before using them in more complex expressions. If students do not answer question 12 and/or question 13 correctly, they may have difficulty with question 14, which requires students to work with two $\square$'s on one side of the equation.

Questions 12–13

12. Which of the following is equal to $\square + \square$?

    (A) $\square$

    (B) $1 \cdot \square$

    (C) $2 \cdot \square$

    (D) $2 + \square$

13. Consider each of the following. Is it equal to $\square - \square$?

yes (no)   **a)** $\square$

(yes) no   **b)** $0 \cdot \square$

(yes) no   **c)** 0

yes (no)   **d)** $2 \cdot \square$

Guiding Class Discussion

In addition to the basic facts mentioned above, the teacher may also want to discuss with the students the fact that 12A and 12B are equal.

The practice sheets and suggested activities in the Follow-up Instruction section under the concept **Group or combine like terms** may be helpful for those students having difficulty or needing more practice.

Practice sheet 6, which emphasizes combining terms containing $\square$'s, is especially appropriate for use with questions 12 and 13.

Question 14

14. In each of the following, what number belongs in the $\square$ to make both sides of the equation equal to the same number?

   a) $5 + \square + \square = 20 - 3$ Answer: _6_

   b) $\square + 3 + \square = 7 + 4$ Answer: _4_

   c) $2 \cdot \square - \square + 4 = 13$ Answer: _9_

   d) $3 \cdot \square - 7 + \square = 18 + 7$ Answer: _8_

Guiding Class Discussion

Most of the concepts addressed thus far are necessary to solve the equations in question 14, but combining like terms is especially important here. It can be pointed out to the students that combining like terms before attempting to substitute a number for $\square$ is much easier than substituting a number at the outset. This could be accomplished by having the students simplify each expression in question 14, and then substitute the respective solutions in the original expressions and in the simplified expressions. For example:

$$5 + \boxed{6} + \boxed{6} = 20 - 3 \quad \text{and} \quad 5 + 2 \cdot \boxed{6} = 20 - 3.$$

The student should then compare the left sides of both equations. Also, it is important to stress the fact that in any one equation, $\square$ always stands for the same number.

The practice sheets and suggested activities in the Follow-up Instruction section under the concept **Group or combine like terms** may be helpful for those students having difficulty or needing more practice.

Practice sheet 6, in particular, could be used at this point to address misunderstandings about combining terms containing $\square$'s.

Focus of Question 15

Question 15 assesses the students' understanding of inverse operations and how they affect the variable. It might be helpful if the teacher talks in terms of "undoing." For example, subtraction "undoes" addition.

Question 15

15. Consider each of the following expressions. If □ stands for a number, does the expression equal □ ?

(yes)	no	**a)**	$5 + \square - 5$
(yes)	no	**b)**	$\square + 5 - 5$
(yes)	no	**c)**	$\dfrac{\square}{5} \cdot 5$
(yes)	no	**d)**	$5 \cdot \square \cdot \dfrac{1}{5}$
yes	(no)	**e)**	$\square + 5 \cdot \dfrac{1}{5}$
yes	(no)	**f)**	$5 \cdot \square - 5$

Guiding Class Discussion

Commutative properties should be discussed in conjunction with 15a and 15d. When discussing 15c and 15d, the teacher should emphasize the fact that multiplying by $\frac{1}{5}$ and dividing by 5 give the same results.

If students have difficulty with this concept, manipulatives could be used to help them get a concrete idea of the equivalence of multiplying by $\frac{1}{5}$ and dividing by 5. For example, the teacher could use a set of objects such as toothpicks, pieces of paper, clips, or buttons. If one starts with 10 toothpicks and models the problem $10 \div 5$ by separating 10 toothpicks into groups of 5, the result is 2. The problem $10 \times \frac{1}{5}$ yields the same result as $10 \div 5$ since $10 \times \frac{1}{5}$ is the same as $\frac{1}{5} \times 10$ and taking $\frac{1}{5}$ of a group of 10 toothpicks yields a result of 2.

The teacher should be alert for students who answer 15e or 15f incorrectly, and discuss with these students which operations undo each other.

The practice sheets and suggested activities in the Follow-up Instruction section under the concepts **Apply the properties of commutativity, associativity, and distributivity** and **Select appropriate inverse operations** may be helpful for those students having difficulty or needing more practice.

Focus of Questions 16–28

The intent of questions 16–28 is to develop the concept of a balanced equation and to show how to maintain that balance when performing arithmetic operations on both sides of the equation. Also, operations on inequalities that parallel those used on equations are stressed. Since this unit does not assume students' familiarity with negative numbers, negative numbers are not used.

Questions 16–17

16.

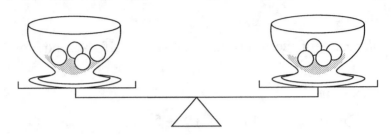

On the scale above, there are 4 balls of equal weight in each bowl. Consider each of the following changes. Would it keep the sides of the scale balanced?

yes	(no)	**a)**	Removing 1 ball from the left side
yes	(no)	**b)**	Adding 1 ball of the same weight to the right side
(yes)	no	**c)**	Removing 1 ball from each side
(yes)	no	**d)**	Adding 2 balls of equal weight to each side
yes	(no)	**e)**	Adding 1 ball to the left side and removing 1 ball of the same weight from the right side
(yes)	no	**f)**	Doubling the number of balls on both sides

17.

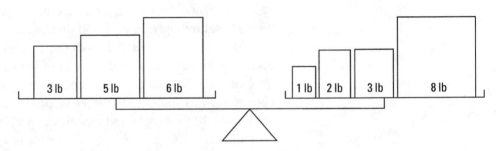

If the 5-pound weight is removed from the left side of this balanced scale, which weight or weights must be removed from the right side to keep it balanced?

Answer: *Remove the 2 lb. and 3 lb. weights from the right side.*

Analysis of Questions 16–17

Questions 16 and 17 use a physical metaphor for a balanced equation. If students have difficulty with question 16, the teacher can use an actual balance-type scale to illustrate each part of the question. The stress should be on doing the same thing to both sides of the scale to keep it balanced.

If students answer 16e incorrectly, it may be because they think they have to undo an operation on one side of an equation by doing the opposite of the operation on the other side. The actual balance can be helpful in showing why this does not apply to opposite sides of the equation.

When discussing 16f, the teacher might also want to discuss the effect of halving the number of balls on each side in anticipation of using fractions.

The purpose of question 17 is to assess whether the students understand that it is the total weight that is important in the balance, not the number of objects. If they grasp this principle, then their understanding may carry over to a more abstract equation such as $3x + 2y + x + 5 = 3y + 10 + 2x$.

Guiding Class Discussion

The practice sheets and suggested activities in the Follow-up Instruction section under the concept **Perform the same operations on both sides of an equation or inequality** may be helpful for those students having difficulty or needing more practice.

Question 18

18. Consider each of the following actions. If 5 is first subtracted from the right side of the given equation, does the action keep the equation balanced?

$$8 + 3 + 2 + 1 = 6 + 5 + 3$$

(yes)	no	**a)**	Subtracting both 3 and 2 from the left side
(yes)	no	**b)**	Subtracting 5 from the left side
yes	(no)	**c)**	Subtracting 3 from the left side
yes	(no)	**d)**	Subtracting both 2 and 1 from the left side

Guiding Class Discussion

In question 18 the balanced scale in question 17 has been translated into a numerical equation. The numbers in this question are identical to those in 17 so that the teacher can relate back to the scale if students have difficulty. (Note, however, that although the numbers are the same in both questions, they appear on different sides of the equation. The teacher may need to remind or reassure the students that the two equations are equivalent.)

A discussion with the class might be helpful to discover how students interpret "subtracting 5." If they think in terms of "removing a 5 that is there," they may answer 18a correctly because "removing both a 3 and a 2" will maintain the balance. For those who think in terms of "subtracting a new 5," 18b will be the "correct" way to maintain the balance. It should be pointed out that both interpretations are correct.

If students answer 18c incorrectly, they may believe that the position of the number is important rather than its value. If they answer 18d incorrectly, they may believe that the number of terms on each side is important for maintaining balance.

The practice sheets and suggested activities in the Follow-up Instruction section under the concept **Perform the same operations on both sides of an equation or inequality** may be helpful for those students having difficulty or needing more practice.

Focus of Question 19

Question 19 assesses the students' understanding of the principles in questions 16–18, but in the context of an inequality.

Question 19

19. This question refers to the following inequality:

$$6 + 4 + 8 < 6 + 4 + 1 + 5 + 3.$$

(Remember "$\square < \triangle$" means "$\square$ is less than $\triangle$.")

In each of the following, start with the inequality above. Perform the indicated operation and write a new inequality. Then indicate whether the new inequality is true.

	Operation	**New Inequality**	**Is It True?**
a)	Subtract 6 from each side.	$4 + 8 < 4 + 1 + 5 + 3$	(yes) no
b)	Subtract 4 from each side.	$6 + 8 < 6 + 1 + 5 + 3$	(yes) no
c)	Subtract 8 from each side.	$6 + 4 < 6 + 4 + 1$	(yes) no

Guiding Class Discussion

In discussing question 19 with the class, the accent should be on maintaining the "imbalance." A brief discussion to refresh the students' understanding of $<$ and $>$ might be necessary.

If students have difficulty with this question, the teacher could use the physical model of the scale again to show how an unbalanced scale will remain unbalanced in the same direction by either subtracting or adding the same weight on both sides.

Depending on their interpretation of the word "subtract," the form of students' answers may vary. For example, other correct forms for 19c are

$$6 + 4 + 8 - 8 < 6 + 4 + 1 + 5 + 3 - 8, \text{ or}$$

$$6 + 4 < 6 + 4 + 1, \text{ or}$$

$$6 + 4 < 6 + 4 + 1 + 5 + 3 - 8.$$

The teacher should point out that all of these forms are correct.

The practice sheets and suggested activities in the Follow-up Instruction section under the concept **Perform the same operations on both sides of an equation or inequality** may be helpful for those students having difficulty or needing more practice.

In particular, by presenting familiar situations that can be represented by inequalities, **practice sheet 7** can help students understand how to maintain the "imbalance" in an inequality.

Focus of Question 20

Question 20 assesses students' understanding of the rules that govern operations on inequalities.

Question 20

20. This question refers to the inequality $10 < 12$.

In each of the following, perform the indicated operation on the inequality above and write a new inequality. Indicate whether the new inequality that is formed is true.

	Operation	New Inequality	Is It True?
a)	Multiply both sides by 3.	$30 < 36$	(yes) no
b)	Divide both sides by 2.	$5 < 6$	(yes) no
c)	Add the same number to each side.	Ex.: $13 < 15$	(yes) no
d)	Subtract the same number from each side.	Ex.: $6 < 8$	(yes) no
e)	Multiply both sides by 0.	$0 < 0$	yes (no)

Analysis of Question 20

Students can probably see in the specific cases of 20a and 20b that the inequality is maintained. It is important that students be warned against jumping to conclusions when using multiplication and division with inequalities. Specifically, the teacher can point to 15e as a counterexample to a rule that parallels a rule governing equations: *Multiplying both sides of an inequality by the same number does not necessarily maintain the inequality.* Only if students are already familiar with operations with negative numbers should examples of multiplying inequalities by negative numbers be presented at this time.

Guiding Class Discussion

In answering 20c and 20d, each student may have added or subtracted a different number. The teacher could have students volunteer their version of new inequalities. After examining several of these, the students could be asked to make a general rule about adding or subtracting the same quantity on both sides of an inequality.

The practice sheets and suggested activities in the Follow-up Instruction section under the concept **Perform the same operations on both sides of an equation or inequality** may be helpful for those students having difficulty or needing more practice.

In particular, if the students have difficulty formulating a rule, the activity on **practice sheet 7** might be helpful.

Focus of Question 21   Question 21 assesses the students' understanding of balanced equations in a more abstract setting.

Question 21

21. In the equation $\Box + 23 = 89$, $\Box$ stands for a number.

If 10 is subtracted from the left side of this equation to give $\Box + 23 - 10$, what must be done to the right side to keep the equation balanced?

Answer: *10 must be subtracted from the right side.*

Analysis of Question 21

Students who have correctly answered similar questions without the $\Box$ may be confused by the $\Box$. Some students may determine what number belongs in the $\Box$ before answering this question. It may reassure both those who found the value that belongs in the $\Box$ and those who did not that their answers are the same.

The teacher should be alert for those students who suggest "adding 10" to undo the subtraction and again stress the necessity of doing the same thing on both sides of the equation.

Guiding Class Discussion

The following question would provide a good basis for discussing the principle of maintaining a balanced equation by performing the same operation on both sides.

In the following equation, $\Box$ stands for a number. The equation is true for that number.

$$2 \cdot \Box + 3 = 13$$

Tell whether the new equation obtained would still be true in each of the following cases.

(yes) no	**a)**	If both sides are divided by 2
(yes) no	**b)**	If 3 is subtracted from both sides
(yes) no	**c)**	If both sides are divided by 3
(yes) no	**d)**	If 2 is subtracted from both sides

In explaining a and c above, there should be considerable discussion of the distributive property. How, for example, should we divide $2 \cdot \Box + 3$ by 2? Possible expressions to use include

$$\frac{1}{2}(2 \cdot \Box + 3), \quad (2 \cdot \Box + 3) \div 2, \quad \text{or} \quad \frac{2 \cdot \Box + 3}{2}.$$

Students should be made aware of these alternative approaches.

A question later in the assessment (25) asks the student to decide which one of the operations listed in this activity is the most desirable one to use first in solving the equation.

The practice sheets and suggested activities in the Follow-up Instruction section under the concept **Perform the same operations on both sides of an equation or inequality** may be helpful for those students having difficulty or needing more practice.

Practice sheet 8 is especially appropriate for emphasizing the distributive property. It can also be used to discuss maintaining balance in an equation.

Focus of Questions 22–23

Questions 22 and 23 assess the students' knowledge of inverse operations. Without using the terminology of inverses, this question tries to determine if the students understand the concept of "undoing" an operation.

Questions 22–23

22. In each of the following, an operation has been performed on $\boxed{27}$. What should be done to undo the operation? That is, what can be done to get $\boxed{27}$ alone? Circle your choice from those listed.

a) $\boxed{27} + 5$ Add 5.   (Subtract 5)   Multiply by 5.   Divide by 5.

b) $5 \cdot \boxed{27}$ Add 5.   Subtract 5.   Multiply by 5.   (Divide by 5.)

c) $\boxed{27} - 5$ (Add 5)   Subtract 5.   Multiply by 5.   Divide by 5.

d) $\dfrac{\boxed{27}}{5}$ Add 5.   Subtract 5.   (Multiply by 5.)   Divide by 5.

e) $\dfrac{1}{5} \cdot \boxed{27}$ Add 5.   Subtract 5.   (Multiply by 5.)   Divide by 5.

23. In each of the following, an operation has been performed on $\square$.

Indicate by circling your choice whether one should add a number, subtract a number, multiply by a number, or divide by a number to undo the operation to get $\square$ alone. Then tell which number should be used.

					Number
a)	$6 \cdot \square$	Add	Subtract	Multiply	(Divide) _6_
b)	$4 + \square$	Add	(Subtract)	Multiply	Divide _4_
c)	$\dfrac{\square}{3}$	Add	Subtract	(Multiply)	Divide _3_
d)	$\square - 7$	(Add)	Subtract	Multiply	Divide _7_
e)	$\dfrac{1}{2} \cdot \square$	Add	Subtract	(Multiply)	Divide _2_

Analysis of Questions 22–23

Question 22 is similar to question 15 but requires the students to choose the correct inverse ("undoing") operation. The teacher should be sure to point out the equivalence of 22d and 22e.

In question 23, the students not only have to choose the operation, but also the number. So, in 23a, students may choose division and/or multiplication. The teacher should make sure that they choose the correct number with either of these, that is, multiply by $\frac{1}{6}$ *or* divide by 6.

Guiding Class Discussion

The practice sheets and suggested activities in the Follow-up Instruction section under the concept **Select appropriate inverse operations** may be helpful for those students having difficulty or needing more practice.

In particular, **practice sheet 8** can be used for more practice with inverse operations.

Focus of Question 24

The purpose of question 24 is to lead the students into solving simple linear equations such as $7 + 8x = 23$.

Question 24

24. In each of the following, two different operations are indicated. To get □ alone, two operations have to be performed. Indicate the operations and numbers that should be used to accomplish this. The first one has been done for you.

a) $8 \cdot \square + 7$ First subtract 7 _____, then divide by 8. _____

b) $6 \cdot \square - 3$ First *add 3* _____, then *divide by 6* _____

c) $\dfrac{\square}{5} + 2$ First *subtract 2* _____, then *multiply by 5* _____

d) $\dfrac{1}{2} \cdot \square + 4$ First *subtract 4* _____, then *multiply by 2* _____

Guiding Class Discussion

In discussing this question, the teacher should point out that taking the inverse of, *i.e.* "undoing", the operation involving the numerical term (the one without the □) is always the first step in solving the equation. If students have difficulty with this idea, encourage them to think of the variable term as one number. For example, $8 \cdot \square + 7$ should be thought of as "a number plus 7." In solving an equation such as $8 \cdot \square + 7 = 55$, the student first would think "a number plus $7 = 55$" and then "8 times a number $= 48$." The teacher should check with the students who chose multiplication for the second blank in b to see that they chose the correct corresponding number. Students should be made aware that multiplication could also be used in part a.

The practice sheets and suggested activities in the Follow-up Instruction section under the concept **Select appropriate inverse operations** may be helpful for those students having difficulty or needing more practice.

Focus of Questions 25–26

Questions 25 and 26 focus on the strategy for solving equations by asking students to determine the best choice for the first operation that should be performed to isolate the variable.

Question 25

25. Suppose we want to get □ by itself on the left side of the equation

$$2 \cdot \square + 3 = 13.$$

Which of the following would be the best choice for the first operation to perform on both sides of the equation?

    (A)   Divide both sides by 2.
    (B)   Subtract 3 from both sides.
    (C)   Divide both sides by 3.
    (D)   Subtract 2 from both sides.

Which would be the best choice for the second step?

    Answer: _A_

Guiding Class Discussion

Question 25 recalls the example from the class discussion of question 21. In this question, strategy should be stressed rather than the actual computations. The teacher can refer back to the class discussion under question 21, if that seems helpful. The teacher can point out that 25A is a possible first step, but that it would introduce fractions into the equation. Students usually prefer to work with whole numbers, but demonstrating a solution that starts with 25A might be beneficial.

The practice sheets and suggested activities in the Follow-up Instruction section under the concepts **Select appropriate inverse operations** and **Perform the same operations on both sides of an equation or inequality** may be helpful for those students having difficulty or needing more practice.

Question 26

26. $$7 \cdot \square + 5 = 35$$

In the equation above, $\square$ stands for a number. Carolyn wanted to find the number that makes the equation true. If she solved the equation correctly, which of the following could have been the result of her first step?

    (A) $\square + 5 = 28$
    (B) $7 \cdot \square = 40$
    (C) $\square + 5 = 5$
    (D) $7 \cdot \square = 30$
    (E) $12 \cdot \square = 35$

Analysis of Question 26

If students choose 26A, they may have incorrectly subtracted 7 from both sides. A choice of 26B may mean they have subtracted 5 on the left side, but added 5 on the right. A choice of 26C may mean they have divided both sides by 7, but did not distribute it properly. A choice of 26E means they have incorrectly combined unlike terms.

Guiding Class Discussion

The following questions could be the basis for an introductory discussion of solving simple linear equations and inequalities in algebra, if the teacher feels the class is ready to extend the ideas presented thus far in the unit.

$$3x - 8 = 21$$

In the equation above, x stands for a number. Carlos solved the equation correctly. Which of the following could have been the result of his first step?

    (A) $x - 8 = 18$
    (B) $x - 8 = 7$
    (C) $3x = 29$
    (D) $3x = 13$
    (E) $5x = 21$

$$5x + 2 > 20$$

Bob solved the inequality above correctly. Which of the following could have been the result of his first step?

    (A) $x + 2 > 4$
    (B) $5x > 18$
    (C) $7x > 20$
    (D) $5x > 22$
    (E) $x + 2 > 15$

The practice sheets and suggested activities in the Follow-up Instruction section under the concepts **Select appropriate inverse** operations and **Perform the same operations on both sides of an equation or inequality** may be helpful for those students having difficulty or needing more practice.

Focus of Question 27

The intent of question 27 is to assess the students' knowledge of the distributive property and their ability to work with it in an equation.

Question 27

27.
$$2(\square + 3) = 6$$

In the equation above, $\square$ stands for a number. Jane has solved the equation correctly. Consider each of the following. Could it have been the result of her first step?

yes (no)　　**a)**　$2 \cdot \square + 3 = 6$

(yes) no　　**b)**　$2 \cdot \square + 2 \cdot 3 = 6$

yes (no)　　**c)**　$\square + 2 \cdot 3 = 6$

(yes) no　　**d)**　$\dfrac{2(\square + 3)}{2} = \dfrac{6}{2}$

yes (no)　　**e)**　$2 \cdot \square + 2 \cdot 3 = 2 \cdot 6$

(yes) no　　**f)**　$\square + 3 = \dfrac{6}{2}$

(yes) no　　**g)**　$\square + 3 = 3$

Analysis of Question 27

Here it is possible to illustrate two equally valid strategies for solving the equation. Some students may see only one correct way of dealing with the parentheses, as in 27b, for example. They will benefit by seeing a variety of ways of working with parentheses in the context of an equation. The teacher should point out how 27d or, alternatively, 27f is just as valid.

Those students who answer 27a and 27c incorrectly are not distributing properly. And those who answer 27e incorrectly may be confused about the distributive property in the context of an equation. The teacher should clarify that the 2 appears only on the left side.

Guiding Class Discussion

The practice sheets and suggested activities in the Follow-up Instruction section under the concepts **Apply the properties of commutativity, associativity, and distributivity** and **Perform the same operations on both sides of an equation or inequality** may be helpful for those students having difficulty or needing more practice. Specifically, the questions on **practice sheet 8** are like assessment question 27. They may be used for practice with the distributive property and/or for discussing strategies for solving equations.

Focus of Question 28

Question 28 presents an equation containing a □ , correctly solved. It summarizes much of the material that preceded it in this unit. Correct answers to this question indicate an understanding of many of the concepts assessed thus far.

Question 28

28.
$$7 \cdot \square - 3 \cdot \square + 5 = 40 - 7$$

The following steps show a correct solution for the equation above. In the space to the right of each step, indicate what was done to get to that step.

a) $7 \cdot \square - 3 \cdot \square + 5 = 33$ Answer: *combined 40 - 7*

b) $4 \cdot \square + 5 = 33$ Answer: *combined 7 · □ - 3 · □*

c) $4 \cdot \square = 28$ Answer: *subtracted 5 from both sides*

d) $\square = 7$ Answer: *divided both sides by 4*

Guiding Class Discussion

If students are successful with question 28, the teacher might want to present an equation containing an x as the variable. The following activity provides an example.

$$5x - 3x + 4 = 10 + 6$$

The following steps show a correct solution for the equation above. In the space to the right of each step, indicate what was done to get that step.

a) $5x - 3x + 4 = 16$ *combined 10 + 6*

b) $2x + 4 = 16$ *combined 5x - 3x*

c) $2x = 12$ *subtracted 4 from both sides*

d) $x = 6$ *divided both sides by 2*

The practice sheets and suggested activities in the Follow-up Instruction section under the concepts **Group or combine like terms, Apply the rules that govern the order of operations, Select appropriate inverse operations,** and **Perform the same operations on both sides of an equation or inequality** may be helpful for those students having difficulty or needing more practice.

Follow-up Instruction

After the Instructional Assessment has been used to pinpoint students' conceptual weaknesses, the teacher may wish to use the following suggested activities and practice sheets to help correct those misconceptions.

Group or combine like terms

Practice sheet 1 can be used to convince students that the various incorrect ways of combining unlike terms in question 4 are indeed incorrect. After numbers are chosen to be substituted for $\square$ and $\triangle$, students can evaluate each of the expressions 1–7 and compare the results with the value of the original expression. The teacher can decide how many columns on the chart the students need to complete to reinforce the correct method for combining like terms. The teacher can choose a pair of numbers for the last column or allow the students to choose their own. If calculators are available and the students are proficient with them, this would be a good activity in which to use them.

It may also be helpful to have the students explain why each of the expressions is or is not equivalent to the original expression.

Practice sheet 6 emphasizes combining terms containing $\square$'s. The first two questions can be used to help students understand why in any one equation $\square$ always represents the same number and why $\square + \square = 2 \cdot \square$. Note that question 3 has additional expressions in column B that resemble those in column A, but are not equivalent to any in column A. The teacher could make up more exercises of a similar nature.

Questions 1, 2, 9, and 23 in the Follow-up Assessment also address this concept.

Apply the rules that govern the order of operations

Activity. Each student could make up a word problem in which the operations of addition and subtraction are necessary. They should then write an expression that symbolizes the concrete problem. The students could exchange their papers and write alternate expressions that are equivalent to the original.

Practice sheet 2 can be used to have the students substitute a value for $\square$ in each part of question 5 and evaluate each expression. Students may have difficulty with part e, especially if they are not familiar with negative numbers. The teacher might allow students to leave this line blank rather than discussing "3 − 4" at this point. They should at least be convinced that the expression is not equivalent to the given one.

The teacher should have available both 4-function and scientific calculators for **practice sheet 3**. For the expression in question 1, the scientific calculator will give the correct result, 11, while the 4-function calculator will give an incorrect result, 21, if the expression is simply entered in order from left to right.

The teacher can use this difference in results to discuss and re-inforce the correct order of operations. The scientific calculator is programmed to do multiplication or division first before addition or subtraction, while the 4-function calculator does the operations in the order in which they are entered.

For question 2 the students enter the expression and observe what happens after each entry. A discussion can ensue about how the entering of a + on the calculator signals the completion of the previous multi-plication because the calculator is programmed to order the operations correctly.

Students who do understand the rules for the order of operations as they apply to numbers may have adopted a different set of rules when □ appears in an expression. **Practice sheet 4** may help them to discover that the rules for numbers still apply.

Practice sheet 5 could be used to see if students have difficulty when the □ is moved around in the equation. Encourage them to answer the questions without first solving for □.

Questions 3–8 and 23 in the Follow-up Assessment also address this concept.

| Apply the properties of commutativity, associativity, and distributivity | **Practice sheets 2–5** (described under the concept Apply the rules that govern the order of operations) and **practice sheet 6** (described under the concept Group or combine like terms) may be helpful in reinforcing this concept as well. |

Practice sheet 8 gives students additional practice in dealing with parentheses and applying the distributive property in the context of an equation.

Questions 2, 3, 5–8, 10, and 22 in the Follow-up Assessment also address this concept.

Select appropriate inverse operations

Practice sheet 8 may also be used to show that in some equations it is convenient first to divide rather than to apply the distributive property. This approach emphasizes that the quantity in parentheses represents a number. For example, $2(\Box + 1) = 10$ would be thought of as "2 times a number equals 10." Then division, the inverse of multiplication, can be used to begin the solution process by "undoing" the multiplication. The teacher may wish to have students compare and contrast the two approaches (division and the distributive property) and to provide ex-amples of when each approach is most convenient.

Question 10, 17–21, and 23 in the Follow-up Assessment also address this concept.

Perform the same operations on both sides of an equation or inequality

The students should notice a pattern in the answers for each question on **practice sheet 7**. To help relate this activity to assessment question 20, the teacher could write sample inequalities to illustrate each of the practice sheet questions. For example, in question 1, the inequalities could be $5 > 1$ and $5 + 2 > 1 + 2$.

This practice sheet may be adapted for use with equations by making the numbers within each situation the same. Question 1, for example, might read, "John is on the fifth step of his ladder. Frank is on the fifth step of an identical ladder. Who is higher on his ladder? Each boy climbs up two more steps. Which boy is higher now?" The situation could be represented by the equations $5 = 5$ and $5 + 2 = 5 + 2$.

Practice sheet 8 also relates to this concept as do questions 11–16 and 20–23 in the Follow-up Assessment.

Bibliography of Resource Materials

Bernard, John E., and Martin P. Cohen. "An Integration of Equation-Solving Methods into a Developmental Learning Sequence." *1988 Yearbook of the National Council of Teachers of Mathematics*: 97–111.

Herscovics, Nicolas, and Carolyn Kieran. "Constructing Meaning for the Concept of Equation." *Mathematics Teacher 73* (November 1980): 572–580.

Kieran, Carolyn. "Two Different Approaches Among Algebra Learners." *1988 Yearbook of the National Council of Teachers of Mathematics*: 91–96.

Whitman, Betsey S. "Intuitive Equation-Solving Skills." *1982 Yearbook of the National Council of Teachers of Mathematics*: 199–204.

Success with Algebra: Equations (Apple II+/IIe/IIc/IIGS, IBM/Tandy 1000, Commodore 64/128). Northbrook, IL: Mindscape, Inc.

Answers

1. a) yes
 b) yes
 c) no
 d) no
2. a) yes
 b) no
 c) no
3. a) no
 b) yes
4. a) yes
 b) no
 c) no
 d) no
 e) no
 f) yes
 g) no
5. a) yes
 b) no
 c) yes
 d) no
 e) no
6. 11
7. a) no
 b) no
 c) no
 d) yes
 e) yes
8. a) yes
 b) yes
 c) yes

 d) yes
 e) yes
9. a) yes
 b) yes
 c) yes
 d) no
10. a) no
 b) yes
 c) no
 d) yes
11. a) 6
 b) 2
 c) 12
 d) 15
12. C
13. a) no
 b) yes
 c) yes
 d) no
14. a) 6
 b) 4
 c) 9
 d) 8
15. a) yes
 b) yes
 c) yes
 d) yes
 e) no
 f) no
16. a) no

 b) no
 c) yes
 d) yes
 e) no
 f) yes
17. 2 lb. and 3 lb.
18. a) yes
 b) yes
 c) no
 d) no
19. a) $4+8 < 4+1+5+3$
 (or $4+8 < 6+4+3$
 or $12 < 13$); yes
 b) $6+8 < 6+1+5+3$
 (or $6+8 < 6+4+5$
 or $14 < 15$); yes
 c) $6+4 < 6+4+1$
 (or $6+4 < 6+5$
 or $10 < 11$); yes
20. a) $30 < 36$; yes
 b) $5 < 6$; yes
 c) Answers will vary;
 for example,
 $13 < 15$; yes
 d) Answers will vary;
 for example,
 $6 < 8$; yes
 e) $0 < 0$; no
21. 10 must be subtracted
 from the right side

22. a) Subtract 5.
 b) Divide by 5.
 c) Add 5.
 d) Multiply by 5.
 e) Multiply by 5.
23. a) Divide; 6
 b) Subtract; 4
 c) Multiply; 3
 d) Add; 7
 e) Multiply; 2
24. b) First add 3,
 then divide by 6.
 c) First subtract 2,
 then multiply by 5.
 d) First subtract 4,
 then multiply by 2.
25. B, A
26. D
27. a) no
 b) yes
 c) no
 d) yes
 e) no
 f) yes
 g) yes
28. a) Combined $40-7$.
 b) Combined
 $7 \cdot \square - 3 \cdot \square$.
 c) Subtracted 5
 from both sides.
 d) Divided both
 sides by 4.

Answers to Practice Sheets

Practice sheet 1
45, 16, 90
1. 45, 16, 90
2. 99, 0, 264
3. 79, 10, 194
4. 45, 6, 120
5. 65, 16, 100
6. 45, 16, 90
7. 59, 13, 106
1 and 6

Practice sheet 2
1. a) 5, 7, 8
   b) 3, 5, 6
   c) 5, 7, 8
   d) 3, 1, 0
   e) 3, 5, 6
a and c
2. a) $\triangle + 15 - 10$
   b) $\square - 5 + 3$
3. The student was not correct. $6 + \square - 4$ does not represent the situation because "selling 6 cars" indicates that 6 should be subtracted from $\square$ and "accepting 4 cars as trade-ins" indicates that 4 should be added to $\square$.

Practice sheet 3
1. 4-function calculator: 21
   Scientific calculator: 11
   a) 4-function
   b) scientific
   c) 4-function
   d) scientific
2. 21, 39, 45
3. a) 4-function
   b) scientific
   c) 4-function

Practice sheet 4
1. Answers may vary. For example,
   Value for $\square$: 0, 1, 2, 5, 10
   Corresponding value for given expression: 3, 9, 15, 33, 63
   a) 0, 9, 18, 45, 90
   b) 18, 19, 20, 23, 28
   c) 6, 9, 12, 21, 36
   d) 3, 9, 15, 33, 63
   e) 3, 9, 15, 33, 63
d and e
2. a) yes
   b) no
   c) yes
   d) yes
   e) no

Practice sheet 5
1. a) yes
   b) yes
   c) yes
   d) yes
   e) yes
2. a) 7, 10, 2
   b) 19
   c) 3, 6
   d) 9
   e) $\square + 19 - 9 = 22$
3. Answers may vary. For example,
   $\square + 16 - 12 = 18$

Practice sheet 6
1. a) $5 + 5 + 6$
   b) $2 \cdot 5$ is another name for $5 + 5$.
2. a) $\square + \square + 4$
   b) $2 \cdot \square + 4$
   c) Answers will vary. For example, if 3 is substituted for $\square$,
   $3 + 3 + 4 = 10$
   $2 \cdot 3 + 4 = 10$
3. a) $4 + 2 \cdot \square$
   b) $2 \cdot \square + 7$
   c) $\square + 3$
   d) $2 \cdot \square - 5$
   e) $3 \cdot \square + 6$
   f) $3 \cdot \square + 7$

Practice sheet 7
1. Frank, Frank
2. the tape, the tape
3. the car, the car
4. Anita's, Anita's
5. Yardley, Yardley
6. B, B
7. The Green Sox, The Green Sox

Practice sheet 8
Answers may vary.
1. $2 \cdot \square + 2 = 10$
   or $\square + 1 = 5$
2. $12 + 3 \cdot \square = 15$
   or $4 + \square = \frac{15}{3}$
3. $4 \cdot \square - 4 \cdot 2 = 12$
   or $\square - 2 = 3$
4. $6 \cdot \square + 9 = 21$
   or $2 \cdot \square + 3 = 7$
5. $15 \cdot \square - 5 = 10$
   or $3 \cdot \square - 1 = 2$
6. $30 - 8 \cdot \square = 6$
   or $15 - 4 \cdot \square = 3$
7. $24 = 4 \cdot \square + 4$
   or $6 = \square + 1$
8. $35 = 20 - 5 \cdot \square$
   or $7 = 4 - \square$
9. $6 \cdot \square + 6 = 12$
   or $2 \cdot \square + 2 = 4$
6. $42 + 6 \cdot \square = 60$
   or $21 + 3 \cdot \square = 30$

Answers to Follow-up Assessment

1. a) yes
  b) no
  c) yes
  d) no
2. a) yes
  b) no
  c) no
  d) no
  e) no
  f) no
3. a) yes
  b) no
  c) yes
  d) no
  e) no
4. 25
5. a) no
  b) no
  c) no
  d) yes
  e) yes
6. a) yes
  b) yes
  c) no
  d) no

  e) yes
7. a) yes
  b) no
  c) yes
  d) no
8. a) 5
  b) 6
  c) 13
  d) 11
9. a) 8
  b) 5
  c) 8
  d) 5
10. a) yes
  b) yes
  c) yes
  d) no
  e) yes
  f) no
11. a) no
  b) no
  c) yes
  d) yes
  e) no
  f) yes

12. 7 lb. and 1 lb.
13. a) yes
  b) yes
  c) no
  d) no
14. a) $3+7 < 3+2+4+6$; yes
  b) $5+7 < 5+2+4+6$; yes
  c) $5+3 < 5+2+6$; yes (or $5+3 < 3+4+6$)
15. a) $18 < 30$; yes
  b) $3 < 5$; yes
  c) $16 < 22$; yes
  d) For example, $4 < 10$; yes
  e) $0 < 0$; no
16. 16 must be added to the right side.
17. a) Add 6.
  b) Divide by 6.
  c) Multiply by 6.
  d) Subtract 6.
  e) Multiply by 6.

18. a) Divide, 3
  b) Add, 9
  b) Multiply, 4
  d) Subtract, 5
  e) Multiply, 6
19. b) subtract 7, divide by 6
  a) add 2, multiply by 3
  d) subtract 9, multiply by 2
20. D, A
21. D
22. a) yes
  b) no
  c) no
  d) yes
  e) yes
  f) no
  g) yes
23. a) Combined $41 - 4$.
  b) Combined $8 \cdot \square - 3 \cdot \square$.
  d) Subtracted 7 from both sides.
  d) Divided both sides by 5.

Glossary

Words, Phrases and Symbols Used in Algebridge™

Working definitions of the words, phrases and symbols that are used in *Algebridge* are given here. This glossary is intended to be a reference for students to use not only as they work with the *Algebridge* materials, but also as they work on other mathematics activities. Spaces are provided for students to write new words, phrases, and symbols that are not included in the glossary.

This glossary is copyrighted by Center for Applied Linguistics, 1118 22nd Street, NW, Washington, DC, 20037, in conjunction with Northern Virginia Community College, Metropolitian State College, and Miami-Dade Community College, and is reprinted by permission. Funding for the glossary was provided by the Fund for the Improvement of Postsecondary Education, U.S. Department of Education.

Common Words and Phrases

Used in Word Problems

Note: Special thanks to the staff of the Mathematics Department of Northern Virginia Community College for this section.

Common Words and Phrases Used in Word Problems

Word or phrase	What you should do	Algebraic expression
Dan's *age* 10 years *ago*	subtract	$a - 10$ (where a is Dan's present age)
Emma's *age* 4 years *from now*	add	$a + 4$ (where a is Emma's present age)
A number *decreased* by 7	subtract	$x - 7$
Five *decreased* by the unknown	subtract	$5 - x$
Difference of a number and 5	subtract	$x - 5$
Distance traveled in t hours at 50 *mph* (miles per hour)	multiply	$50t$
Distance traveled in 40 *minutes* at r *mph* (40 minutes = 2/3 hour)	multiply	$(2/3)r$ or $2r/3$
Two *less than* the unknown	subtract	$x - 2$
Two *less* the unknown	subtract	$2 - x$
Three *less than* the *cube of b*	subtract	$b^3 - 3$
Five *less than* the *product* of x and y	subtract	$xy - 5$

Word or phrase	What you should do	Algebraic expression
Miles per gallon if 3 gallons are used in x miles	divide	$x/3$
Nine **minus** the product of x and y	subtract	$9 - xy$
Three is four **more than** a number	add	$3 = x + 4$
Five **more than** the unknown	add	$x + 5$
Three **more than** twice the unknown	add	$2x + 3$
Two **more than** the square of x	add	$x^2 + 2$
40 **percent mark up** on an item	multiply and then add	$x + 0.4x$, or $1.4x$
40 **percent off** the purchase price (p)	multiply and then subtract	$p - 0.4p$, or $0.6p$
10 **percent off** a sale item that is already marked down 30%	multiply	$(0.9)(0.7)(x)$
Product of a number **and 5**	multiply	$5x$
Quotient of a number **and 5**	divide	$x/5$
Rate when traveling **with** the wind (w)	add	$r + w$
Rate when traveling **against** the wind (w)	subtract	$r - w$

Word or phrase	What you should do	Algebraic expression
Rate when traveling x miles in 3 **hours**	divide	$x/3$
Separate 17 into two parts	subtract	x and $17 - x$
Sum of a number **and 5**	add	$x + 5$
Sum of two times x and three times y	add	$2x + 3y$
Sum of the squares of a and b	add	$a^2 + b^2$
Time in hours when traveling 4 miles at **rate r** (miles per hour)	divide	$4/r$
Four **times as much**	multiply	$4x$
Five **times** x divided by y	multiply	$5x/y$
Six **times** w **less** 4	mulitply and subtract	$6w - 4$
Four **times** the **sum** of x and 2	add and then multiply	$4(x + 2)$
Six **times** the quantity w **minus** 4	subtract and then multiply	$6(w - 4)$
Twice the unknown	multiply	$2x$
Twice Maria's age 3 years **ago**	subtract and then multiply	$2(a - 3)$ (where a is Maria's present age)

Use the chart below to fill in other words or phrases you have found
in word problems that you want to remember.

Word or phrase	What you should do	Algebraic expression

Mathematical Symbols and
Their Meanings

Addition

Symbol	Related words	Examples	Explanation
	add addition	3 + 4	"3 + 4" means *add* 3 and 4; do the operation of *addition*
+	plus increased by more than	"3 + 4" may be read "3 *plus* 4" or "3 *increased by* 4" or "4 *more than* 3."	The symbol for addition is the *plus* sign: +
	sum	Find the *sum* 3 plus 4.	The answer to an addition problem is called *sum* of the two numbers. Caution: Do not confuse the words *sum* (requiring addition) with the word *product* (requiring multiplication).
+	positive	+2	The plus sign is also used to contrast a *positive* number with a *negative* number, as in the statement "+2 is larger than −5." However, a positive number does not need to have a "+" sign; numbers without any sign such as 3 or 26 or 78 are also known as positive numbers.

Subtraction

Symbol	Related words	Examples	Explanation
	subtract (from) subtraction	5 – 3	"5 – 3" means *subtract* 3 *from* 5; do the operation of *subtraction*
–	minus decreased by less than	"5 – 3" may be read "5 *minus* 3" or "5 *subtract* 3" or "5 *decreased by* 3" or "3 *less than* 5" or (infrequently) "5 *less* 3."	The symbol for subtraction is the *minus* sign: –
	difference	Find the *difference* of 5 minus 3.	The answer to a subtraction problem is called the *difference*.
–	minus negative additive inverse of opposite of	–3	The minus sign is also used as part of the name of a negative number (as in –3). The symbol –3 may be read as "*minus* 3" or "*negative* 3" or "the *additive inverse of* 3" or "the *opposite of* 3."

Multiplication

Symbol	Related words	Examples	Explanation
	multiply multiplication times multiplied by multiplier factor	"2 · 5" may be read "2 *times* 5" or "2 *multiplied by* 5."	"2 · 5" means *multiply*, do the operation of *multiplication*. The 2 and the 5 are called *multipliers* or *factors*.
			The symbols for multiplication are:
·		<u>2 · 5</u>	a dot (·) as in 2 · 5 Note: The dot is above the writing line, whereas in decimal numbers, like 9.38, the decimal point is *on* the writing line.
×		2 × 5	a small ×, as in 2 × 5
( )( ) [ ] { }	parentheses brackets braces	(2)(5) 2(5) (2)5	one or two pairs of parentheses, as in (2)(5). Multiply each term inside the parentheses. Note: There is no symbol between the pairs of parentheses. Braces and brackets are used like parentheses.
		xy $3xy$	two or more letters, or a number and one or more letters next to each other with no space in between.
	product	Find the *product* of 5 times 2.	The answer to a multiplication problem is called a *product*. Caution: Do not confuse the word *product* (requiring multiplication) with the word *sum* (requiring addition).

Multiplication (continued)

Symbol	Related words	Examples	Explanation
	of	1/2 of 10	The word *of* sometimes means multiplication, as in "one half *of* ten is five," which means "multiply one-half times ten to get five." Hint: "of" is used for multiplication only with fractions and percents.
	3 times as large as	$3y$	"Three times as large as y" means multiply 3 times y.
$\times$	twice as large as double twice	$2y$	"Two times as large as y" means multiply 2 times y, and is often expressed as "twice as large as y."
	increased by a factor of	5×3	"Three increased by a factor of five."

Division

Symbol	Related words	Examples	Explanation
	divide division		"12 ÷ 4" means *divide* 12 *by* 4, or divide 4 *into* 12; do the operation of *division*. The symbols for division are:
÷	divided by	12 ÷ 4 is read "12 *divided by* 4."	a "÷", as in 12 ÷ 4
) ‾	divided into	$4)\overline{12}$ is read "4 *divided into* 12."	a ")‾" Caution: Do not confuse this sign with the *radical sign* ($\sqrt{\ }$) for finding roots.
a/b or $\dfrac{a}{b}$	over	12/4 or $\dfrac{12}{4}$ is read "12 *over* 4" or "12 *divided by* 4."	$\dfrac{a}{b}$ and a/b Note: These can be thought of as a division problem (3/4 = .75) or as a fraction (for example, the fraction 3/4).
	divide equally among	"*Divide* 12 cookies *equally among* 4 children."	Separate into equal sized parts.

Equations

Symbol	Related words	Examples	Explanation
=	equals equal sign is equal to	$3 + 2 = 5$	The symbol "$=$" is called the *equal sign*, but it is read "*equals*" (note the final *s*) or "*is equal to.*"
	same identical	$3 + 2 = 5$ is read "3 plus 2 *equals* 5" or "3 plus 2 *is equal to* 5."	*Equal* means *same* or *identical*.
	equality equation	$42 + 8 = 50$	A sentence that contains an equal sign is called an *equation*.

Inequalities

Symbol	Related words	Examples	Explanation
$<$	is less than	$3 < 5$	This is read "3 *is less than* 5." *Less than* means *smaller than* or *to the left of* on the number line.
		←+++|+++++→ -3 -2 -1 0 1 2 3 4 5	Caution: The word *less* is sometimes used for subtraction, so that $6 - 4$ might be read "6 *less* 4" or "4 *less than* 6."
$>$	is greater than	$5 > 3$	This is read "5 *is greater than* 3" or "5 *is more than* 3." *Greater than* means *larger than* or *to the right of* on the number line. $5 > 3$ says the same thing as $3 < 5$.
	is more than	←++++|++++→ -3 -2 -1 0 1 2 3 4 5	Hint: An easy way to remember the symbols for *greater than* and *less than* is to remember that the larger number faces the larger or "open" side of the symbol, for example $75 > 40$ and $40 < 75$.
$\leq$	is less than or equal to	$x \leq 5$	This says that x is *smaller than or equal to* 5.
$\geq$	is greater than or equal to	$x \geq 3$	This says that x is *greater than or equal to* 3. Note: Every statement that includes $<$ or $>$ or $\leq$ or $\geq$ is called an *inequality* (as opposed to an *equation*).

Decimals

Symbol	Related words	Examples	Explanation
.	decimal number	3.4 3.79	Numbers that are not integers may be written as *decimal numbers*. Example 3 and 4/10 or 3 plus 4/10 may be written 3.4; and 3.79 means 3 and 79/100 or 3 plus 79/100.
	decimal point	3.4 is read "three *point* four" or "three and four-tenths." 3.79 is read "three *point* seven nine' or "three and seventy-nine hundredths.	Note: The decimal point is on the line, whereas in multiplication the dot is above the line. Example: 3.4 is a decimal, whereas 3 · 4 means that 3 is multiplied by 4.
,	comma	1,234	Caution: Decimal points and commas are used differently in different languages. Thus, the number one thousand two hundred thirty-four and five-tenths is: 1,234.5 in English 1,234,5 in Arabic 1.234,5 in Spanish, French, etc. Some languages also use apostrophes to separate parts of large numbers. For example, 3'675.471.234,5.

Exponentiation

Symbol	Related words	Examples	Explanation
a^x	exponenti- ation raising to a power	2^3 2^{-5} 2^3 is read as "2 to the 3rd *power*" or "2 to the 3rd"	$2 \cdot 2 \cdot 2$ can be written in a simpler form as 2^3. In this form the operation is called *exponentiation* or *raising to a power*. Remember: a^1 is the same as a.
	base exponent power	2^{-5} is read as "2 to the negative 5th *power*" or "2 to the negative 5"	In 2^3, the 2 is called the *base* and the 3, the small number above and to the right, is called the *exponent* or *power*. Caution: $8^4 = 8 \times 8 \times 8 \times 8$, *not* 8×4
2	squared	7^2 is usually read as "7 *squared*"	*Squared* means a number times itself or a number raised to the second power $(7 \times 7$, or $7^2)$.
3	cubed	7^3 is usually read as "7 *cubed*"	*Cubed* means a number raised to the third power $(7 \times 7 \times 7$, or $7^3)$ or used as a factor three times.

Grouping

Symbol	Related words	Examples	Explanation
( )	parentheses	$7(3 + 4)$ means $7(7)$	These symbols tell you which of the operations come first. Perform the operation inside the parentheses, braces, or brackets first, and then the operation outside the parentheses.
[ ]	brackets		
{ }	braces		

Percent

Symbol	Related words	Examples	Explanation
%	percent	25%	*Percent* is a way of indicating how many parts per hundred. To convert a percent to a fraction, write the percent as the numerator and write 100 as the denominator. Thus, 38% means 38/100. To convert a percent to a decimal number, divide the percent number by 100. Thus, 38% means 38/100, or 0.38.
	of		In percent problems, *of* means *multiply*. Thus, if you own 38% *of* $500, the amount you own is (38%)($500) = (0.38)($500) = $190. Note: *Of* is also used in multiplication of fractions. (See *multiplication*.)
	percentage		The word *percentage* is sometimes used (incorrectly) to mean percent. Strictly speaking, percentage means *the amount in one part*. For example, if you own 38% of $500, your percentage is $190.

Roots

Symbol	Related words	Examples	Explanation
$\sqrt{}$	square root	$\sqrt{9} = 3$	The symbols in the first column indicate the operations of finding roots of numbers. Thus $\sqrt[4]{c}$ means "Find the **4th root** of c."
$\sqrt[3]{}$	cube root	$\sqrt[3]{8} = 2$	
$\sqrt[4]{}$	4th root	$\sqrt{c}$ is usually read "the **square root** of c" rather than "the 2nd root of c."	Note that $\sqrt{c}$ could be written as $\sqrt[2]{c}$ but the 2 is usually omitted.
$\sqrt[5]{}$	5th root		
		$\sqrt[3]{c}$ is usually read "the **cube** root of c" rather than "the 3rd root of c."	
	radical radical sign		The symbol $\sqrt{}$ is called a **radical** or **radical sign**.
	index		The little number attached to the radical sign (for example, the 4 in $\sqrt[4]{c}$ is called the **index**.) Caution: Be careful not to confuse the radical sign ($\sqrt{}$) with the American symbol for division, as in $3\overline{)\,12}$. By the way, the word **root** has a second and completely different meaning in algebra: the **root** of an equation is a number that makes the equation true. (5 is the root of the equation $x + 3 = 8$.)

Practice Sheets

Fractions in Expressions and Equations

Practice Sheet 1

Name: _____

Write each of the following as a fraction in column A. Then write it as another equivalent fraction in column B.

		Column A	Column B
Example	5	$\dfrac{30}{6}$	$\dfrac{5}{1}$
1.	$6 \div 10$		
2.	$5\dfrac{1}{2}$		
3.	$\dfrac{1+2}{13-7}$		
4.	$\dfrac{7}{8} \cdot \dfrac{8}{7}$		
5.	$\dfrac{\frac{2}{3}}{\frac{3}{4}}$		
6.	18		
7.	(Optional) 0.75		
8.	(Optional) 3%		

Fractions in Expressions and Equations

Practice Sheet 2

Name: _____

Match one expression in column A with an expression in column B that is equal to it and write the resulting equation in column C next to the chosen expression from column B.

	Column A	Column B	Column C
Example	$2 + 2$	$2 + \dfrac{1}{2}$	
1.	$2\dfrac{1}{2}$	1	
2.	22	$2 \cdot 2 + 2 \cdot 2$	
3.	$2 + 2 + 2$	4	$2 + 2 = 4$
4.	$2 \cdot (2 + 2)$	$\dfrac{3}{4}$	
5.	$\dfrac{1}{2} + \dfrac{1}{2}$	3	
6.	$\left[\dfrac{1}{2} \div \dfrac{1}{2}\right] + \left[\dfrac{1}{2} \div \dfrac{1}{2}\right]$	$11 + 11$	
7.	$(2 - 2) - (2 - 2)$	$\dfrac{1}{2}$	
8.	$\left[\dfrac{1}{2} \div \dfrac{1}{2}\right] + \left[\dfrac{1}{2} - \dfrac{1}{2}\right]$	6	
9.	$\left[\dfrac{1}{2} \cdot \dfrac{1}{2}\right] + \left[\dfrac{1}{2} \cdot \dfrac{1}{2}\right]$	0	
10.	$\left[2 \div \dfrac{1}{2}\right] - \left[2 \cdot \dfrac{1}{2}\right]$	2	
11.	$\left[\dfrac{1}{2} \cdot 2\right] - \left[\dfrac{1}{2} \div 2\right]$	$\dfrac{2}{2}$	

Fractions in Expressions and Equations

Practice Sheet 3

Name: _____   A L G E |B|R|I|D|G|E|™

Circle the expression(s) in column B that is (are) equal to the expression in column A.

	Column A	Column B
Example	$\dfrac{5}{5}$	$\boxed{5 \div 5}$ or $\dfrac{1}{5} \div 5$ or $5 \div \dfrac{1}{5}$
1.	$\dfrac{6}{1}$	$1 \div 6$ or $6 \div 1$ or $1 \div \dfrac{1}{6}$
2.	$\dfrac{1}{8}$	$1 \div 8$ or $1 \div \dfrac{1}{8}$ or $8 \div 1$
3.	$\dfrac{6}{7}$	$6 \cdot \dfrac{1}{7}$ or $7 \div 6$ or $6 \div 7$
4.	$\dfrac{10}{9}$	$9 \div 10$ or $10 \div 9$ or $9 \cdot \dfrac{1}{10}$
5.	$\dfrac{4}{5}$	$4 \div 5$ or $5 \div 4$ or $5 \div \dfrac{1}{4}$
6.	$\dfrac{8}{9}$	$9 \div 8$ or $9 \cdot \dfrac{1}{8}$ or $9{\overline{)8}}$
7.	$\dfrac{3}{2}$	$2{\overline{)3}}$ or $3 \div 2$ or $3 \cdot \dfrac{1}{2}$

Fractions in Expressions and Equations

Practice Sheet 4

Name: _____

Answer each question in the space provided.

1. Choose a number. Write at least 5 other expressions (names) for the number you choose.

   Example

   Number chosen _____ 17 _____

   _____ $8 + 9$ _____

   _____ $34/2$ _____

   _____ 1×17 _____

   _____ $18 - 1$ _____

   _____ $\sqrt{289}$ _____

   Number chosen _____

   a) _____

   b) _____

   c) _____

   d) _____

   e) _____

2. Choose another number. Write 5 other expressions (names) for the number you choose. The expressions should involve more than one operation.

   Example

   Number chosen _____ 53 _____

   _____ $100 \div 2 + 3$ _____

   _____ $26 \times 2 + 1$ _____

   _____ $40 \times 2 - 27$ _____

   _____ $\sqrt{100} \times 6 - 7$ _____

   _____ $15 \times 3 \div 5 \times 6 - 1$ _____

   Number chosen _____

   a) _____

   b) _____

   c) _____

   d) _____

   e) _____

Fractions in Expressions and Equations

Practice Sheet 5

Name: _____ ALGE|B|R|I|D|G|E|™

In column B use the symbols $=$, $>$, or $<$ to express the relationship between the terms above and below the fraction bar. Then check whether the expression in column A is equal to or is not equal to 1.

	Column A	Column B	Expression in Column A is equal to 1	Expression in Column A is not equal to 1
Example	$\dfrac{7+2}{2+7}$	$7+2\ \boxed{=}\ 2+7$	√	
1.	$\dfrac{78-69}{69-78}$	$78-69\ \boxed{}\ 69-78$		
2.	$\dfrac{45\cdot 6}{6\cdot 45}$	$45\cdot 6\ \boxed{}\ 6\cdot 45$		
3.	$\dfrac{8+9}{8-9}$	$8+9\ \boxed{}\ 8-9$		
4.	$\dfrac{10\div 6}{6\div 10}$	$10\div 6\ \boxed{}\ 6\div 10$		
5.	$\dfrac{5\cdot 9}{9\div 5}$	$5\cdot 9\ \boxed{}\ 9\div 5$		
6.	$\dfrac{\frac{1}{2}+\frac{1}{3}}{\frac{1}{3}+\frac{1}{2}}$	$\frac{1}{2}+\frac{1}{3}\ \boxed{}\ \frac{1}{3}+\frac{1}{2}$		
7.	$\dfrac{\frac{1}{2}\cdot\frac{3}{2}}{\frac{1}{2}\div\frac{2}{3}}$	$\frac{1}{2}\cdot\frac{3}{2}\ \boxed{}\ \frac{1}{2}\div\frac{2}{3}$		
8.	$\dfrac{\triangle-6}{\triangle+6}$	$\triangle-6\ \boxed{}\ \triangle+6$		

Fractions in Expressions and Equations

Practice Sheet 6

Name: _____ A L G E B R I D G E™

Decide whether the number in the left column is greater than, equal to, or less than 1 and place a check in the column that indicates your choice.

	Number	Greater than 1	Equal to 1	Less than 1
Example	$\frac{2}{1}$	√		
1.	$\frac{1}{2}$			
2.	$\frac{2}{2}$			
3.	$\frac{4}{2}$			
4.	$\frac{5}{6}$			
5.	$\frac{6}{6}$			
6.	$\frac{7}{6}$			
7.	$\frac{\frac{1}{4}}{\frac{1}{4}}$			
8.	$\frac{\frac{1}{4}}{\frac{1}{3}}$			
9.	$\frac{\frac{1}{4}}{\frac{1}{5}}$			
10.	$\frac{\frac{1}{5}}{2}$			

Fractions in Expressions and Equations

Practice Sheet 7

Name: _____ ALGE|B|R|I|D|G|E|™

Decide whether the expression in the left column is greater than, equal to, or less than 17 and place a check in the column that indicates your choice.

	Expression	Greater than 17	Equal to 17	Less than 17
Example	$17 \cdot 0$			✓
1.	$17 \cdot 1$			
2.	$\dfrac{17}{1}$			
3.	$17 \cdot \dfrac{6}{6}$			
4.	$\dfrac{17}{5} \cdot 5$			
5.	$17 \cdot \dfrac{8}{9}$			
6.	$17 \cdot \dfrac{11}{9}$			
7.	$\dfrac{17}{2}$			
8.	$\dfrac{1}{17}$			
9.	$\dfrac{3}{17}$			
10.	$\dfrac{5}{17} \cdot 17$			

Fractions in Expressions and Equations

Practice Sheet 8

Name: _____

Simplify each of the following complex fractions. Compare your answers with those of other students and discuss various ways to show that the two fractions are equal.

Procedure: Find the least common denominator (LCD) of all of the fractions. Multiply each term by the LCD. Simplify the result if necessary.

Example
$$\dfrac{\dfrac{1}{5}}{\dfrac{3}{10}} = \dfrac{\dfrac{1}{5}\cdot 10}{\dfrac{3}{10}\cdot 10} = \dfrac{2}{3}$$

1. $\dfrac{\dfrac{1}{3}}{\dfrac{1}{7}} =$

2. $\dfrac{\dfrac{2}{6}}{\dfrac{3}{5}} =$

3. $\dfrac{\dfrac{7}{4}}{\dfrac{1}{6}} =$

Example
$$\dfrac{\dfrac{1}{2}+\dfrac{1}{3}}{\dfrac{1}{2}-\dfrac{1}{3}} = \dfrac{\dfrac{1}{2}\cdot 6+\dfrac{1}{3}\cdot 6}{\dfrac{1}{2}\cdot 6-\dfrac{1}{3}\cdot 6} = \dfrac{3+2}{3-2} = \dfrac{5}{1} = 5$$

4. $\dfrac{\dfrac{5}{6}+\dfrac{8}{3}}{\dfrac{11}{6}-\dfrac{1}{4}} =$

5. $\dfrac{1-\dfrac{1}{8}}{\dfrac{2}{3}} =$

Fractions in Expressions and Equations

Practice Sheet 9

Name: _____

Each of the equations in column A can be multiplied by many numbers in order to get an equation without fractions. Write one of these numbers in column B. Multiply all the terms in the equation by this number and write the resulting equation in column C.

	Column A	Column B	Column C
Example	$\dfrac{2}{6} + \dfrac{5}{3} = 2$	6	$2 + 10 = 12$
1.	$\dfrac{7}{8} + \dfrac{2}{16} = 1$		
2.	$\dfrac{9}{11} + \dfrac{2}{33} = \dfrac{29}{33}$		
3.	$\dfrac{1}{3} + \dfrac{1}{4} + \dfrac{1}{6} = \dfrac{3}{4}$		
4.	$\dfrac{5}{9} + \dfrac{7}{4} = \dfrac{2}{9} + \dfrac{75}{36}$		
5.	$1 = \dfrac{7}{8} + \dfrac{3}{16} - \dfrac{1}{16}$		
6.	$\dfrac{1}{2} + \dfrac{4}{3} = \dfrac{67}{30} - \dfrac{2}{5}$		
7.	$\dfrac{7}{2} = \dfrac{10}{4} - 5 + \dfrac{36}{6}$		
8.	$\dfrac{19}{21} + 2 = \dfrac{7}{3} + \dfrac{4}{7}$		

Fractions in Expressions and Equations

Practice Sheet 10

Name: _____

For each of the following equations, find a common denominator and multiply each term of the equation by the common denominator to get an equation without fractions.

	Equation with Fractions	Common Denominator	Equation without Fractions
Example	$\dfrac{1}{2} + \dfrac{1}{3} = \dfrac{5}{6}$	6	$3 + 2 = 5$
1.	$\dfrac{2}{5} + \dfrac{1}{2} = \dfrac{9}{10}$		
2.	$\dfrac{6}{12} + \dfrac{5}{18} = \dfrac{7}{9}$		
3.	$\dfrac{1}{3} - \dfrac{2}{7} = \dfrac{3}{7} - \dfrac{8}{21}$		
4.	$1 + \dfrac{5}{8} = \dfrac{3}{8} + \dfrac{3}{4} + \dfrac{1}{2}$		
5.	$\dfrac{2}{3} + \dfrac{5}{9} + \dfrac{7}{6} = 4 - \dfrac{7}{6} - \dfrac{4}{9}$		
6.	$\dfrac{5}{2 \cdot 3} + \dfrac{7}{3 \cdot 3} = \dfrac{29}{2 \cdot 3 \cdot 3}$		
7.	$\dfrac{1}{2 \cdot 2} - \dfrac{2}{3 \cdot 5} = \dfrac{1}{2 \cdot 3} - \dfrac{1}{2 \cdot 2 \cdot 5}$		
8.	$\dfrac{7}{2 \cdot 3 \cdot 5} = \dfrac{1}{2 \cdot 2 \cdot 2} + \dfrac{1}{3 \cdot 3} - \dfrac{11}{2 \cdot 2 \cdot 2 \cdot 3 \cdot 3} + \dfrac{3}{5 \cdot 2 \cdot 2}$		

Fractions in Expressions and Equations

Practice Sheet 11

Name: _____ ALGE|B|R|I|D|G|E|™ _____

All of the steps shown in 1 and 2 below are correct. Describe in words what has been done in each step. Then answer question 3.

1. $\dfrac{1}{3} + \dfrac{1}{6} + \dfrac{1}{8}$

$$= \dfrac{8 \cdot 1}{8 \cdot 3} + \dfrac{4 \cdot 1}{4 \cdot 6} + \dfrac{3 \cdot 1}{3 \cdot 8} \qquad \textbf{a)} \ \underline{\hspace{7cm}}$$

$$= \dfrac{8}{24} + \dfrac{4}{24} + \dfrac{3}{24} \qquad \textbf{b)} \ \underline{\hspace{7cm}}$$

$$= \dfrac{15}{24} \qquad \textbf{c)} \ \underline{\hspace{7cm}}$$

$$= \dfrac{3 \cdot 5}{3 \cdot 8} \qquad \textbf{d)} \ \underline{\hspace{7cm}}$$

$$= \dfrac{5}{8} \qquad \textbf{e)} \ \underline{\hspace{7cm}}$$

2. $\dfrac{1}{3} + \dfrac{1}{6} + \dfrac{1}{8} = \dfrac{5}{8}$

$$24 \cdot \dfrac{1}{3} + 24 \cdot \dfrac{1}{6} + 24 \cdot \dfrac{1}{8} = 24 \cdot \dfrac{5}{8} \qquad \textbf{a)} \ \underline{\hspace{6cm}}$$

$$\dfrac{24}{3} + \dfrac{24}{6} + \dfrac{24}{8} = \dfrac{24 \cdot 5}{8} \qquad \textbf{b)} \ \underline{\hspace{6cm}}$$

$$8 + 4 + 3 = 3 \cdot 5 \qquad \textbf{c)} \ \underline{\hspace{6cm}}$$

$$15 = 15 \qquad \textbf{d)} \ \underline{\hspace{6cm}}$$

3. Explain why the steps are different in questions 1 and 2.

Fractions in Expressions and Equations

Practice Sheet 12

Name: _____

Answer each of the questions in the space provided.

1. Consider each of the following. If 14 is placed in the □, is the statement true? Circle **yes** or **no**. If the answer is no, find a value to place in the □ that will make the statement true.

                                                               Value for   □

    yes   no   **a)** $\dfrac{5}{14} \cdot \square$ is another name for $\dfrac{5}{14}$.   _____

    yes   no   **b)** $\dfrac{5}{7} \cdot \square + \dfrac{6}{14} \cdot \square$ is another name for $\dfrac{5}{7} + \dfrac{6}{14}$.   _____

    yes   no   **c)** $\dfrac{5}{7} \cdot \square + \dfrac{6}{14} \cdot \square$ is another name for $10 + 6$.   _____

    yes   no   **d)** $\dfrac{\frac{2}{14} \cdot \square}{\frac{3}{7} \cdot \square}$ is another name for $\dfrac{\frac{2}{14}}{\frac{3}{7}}$.   _____

    yes   no   **e)** $\dfrac{\frac{2}{14} \cdot \square}{\frac{3}{7}}$ is another name for $\dfrac{\frac{2}{14}}{\frac{3}{7}}$.   _____

2. Consider each of the following. Is it another name for $\dfrac{2}{3} + \dfrac{1}{4}$? Circle yes or no. State whether the expression for adding the two fractions is correct or incorrect and explain your decision.

    yes   no   **a)** $\dfrac{2}{3 \cdot 4} + \dfrac{1}{4 \cdot 3}$ _____

    _____

    yes   no   **b)** $\dfrac{2}{3} \cdot \dfrac{2}{2} + \dfrac{1}{4} \cdot \dfrac{2}{2}$ _____

    _____

    yes   no   **c)** $\dfrac{2}{3} \cdot \dfrac{3}{3} + \dfrac{1}{4} \cdot \dfrac{4}{4}$ _____

    _____

    yes   no   **d)** $\dfrac{2}{3} \cdot \dfrac{4}{4} + \dfrac{1}{4} \cdot \dfrac{3}{3}$ _____

    _____

Fractions in Expressions and Equations

Practice Sheet 13

Name: _____

For each of the questions below, be sure that you state whether the expressions have the same value and that you explain your decision.

1. Consider each of the following. Decide whether the value of the expression in column A is or is not equal to the value of the expression in column B. In the space provided, write **equal** or **not equal** and explain your answer.

Column A	Column B	Explanation
a) $\dfrac{5}{7} + 3$	$5 + 21$	
b) $\dfrac{3}{4} + 7$	$\dfrac{3}{4} + \dfrac{28}{4}$	
c) $6 + \dfrac{3}{5}$	$\dfrac{5 \cdot 6}{5 \cdot 1} + \dfrac{3}{5}$	
d) $\dfrac{3}{2} + \dfrac{5}{3}$	$3 \cdot 3 + 2 \cdot 5$	
e) $\dfrac{4}{7} + \dfrac{2}{6}$	$\dfrac{4 \cdot 6}{7 \cdot 6} + \dfrac{2 \cdot 7}{6 \cdot 7}$	
f) $\dfrac{4}{7} + \dfrac{2}{6}$	$\dfrac{4 + 2}{7 + 6}$	
g) $2 - \dfrac{5}{7}$	$\dfrac{14 \cdot 2}{14 \cdot 1} - \dfrac{2 \cdot 5}{2 \cdot 7}$	

2. Consider each of the following. If the same number is placed in each □, is the value of the expression in column A equal to the value of the corresponding expression in column B? Write **equal** or **not equal** and explain your answer.

	Column A	**Column B**	**Explanation**
a)	$\dfrac{\square}{3} + 4$	$\square + 12$	_____
			_____
b)	$7 + \dfrac{\square}{2}$	$\dfrac{2 \cdot 7}{2 \cdot 1} + \dfrac{\square}{2}$	_____
			_____
c)	$\dfrac{\square}{5} + \dfrac{2}{3}$	$\dfrac{3 \cdot \square}{3 \cdot 5} + \dfrac{5 \cdot 2}{5 \cdot 3}$	_____
			_____
d)	$\dfrac{\square}{3} - \dfrac{5}{3}$	$\dfrac{\square - 5}{3}$	_____
			_____
e)	$\dfrac{\square}{4} + \dfrac{5}{5}$	$\dfrac{\square}{4 \cdot 5} + \dfrac{5}{5 \cdot 4}$	_____
			_____

Fractions in Expressions and Equations

Practice Sheet 14

Name: _____ A L G E $|$**B**$|$**R**$|$**I**$|$**D**$|$**G**$|$**E** _____

Consider each of the following. Is it equal to $\frac{6}{8}$? Circle **yes** or **no**.

yes	no	**1.**	$6\overline{)8}$
yes	no	**2.**	$\frac{3}{4}$
yes	no	**3.**	$6 \div 8$
yes	no	**4.**	$8\overline{)6}$
yes	no	**5.**	$\frac{3 \cdot 2}{4 \cdot 2}$
yes	no	**6.**	$6 \cdot \frac{1}{8}$
yes	no	**7.**	$\frac{8}{6}$
yes	no	**8.**	$8 \div \frac{1}{6}$
yes	no	**9.**	$\frac{24}{32}$
yes	no	**10.**	$\frac{6}{8} \cdot \frac{5}{5}$
yes	no	**11.**	$\frac{6}{8} \cdot 8$
yes	no	**12.**	$\frac{6+4}{8+4}$
yes	no	**13.**	$\frac{6(5+1)}{8(5+1)}$

Fractions in Expressions and Equations

Practice Sheet 15

Name: _____

Answer each question in the space provided.

1. Consider each of the following. Is it an equation obtained by multiplying each term of $\frac{2}{3} + \frac{5}{7} = \frac{29}{21}$ by some number? Circle **yes** or **no**. If your answer is yes, write the number on the line beside the equation.

   yes no **a)** $14 + 15 = 29$ _____

   yes no **b)** $2 + 5 = 29$ _____

   yes no **c)** $\frac{2}{21} + \frac{5}{21} = \frac{7}{21}$ _____

   yes no **d)** $0 + 0 = 0$ _____

   yes no **e)** $28 + 30 = 58$ _____

   yes no **f)** $8 + 7 = 15$ _____

2. If every term in the equation $\frac{5}{8} + \frac{5}{8} + \frac{3}{4} = 2$ is multiplied by 8, the result is which of the following equations? Circle your answer. Explain why each of the other choices is incorrect.

   (A) $\frac{5}{8} + \frac{5}{8} + \frac{6}{8} = \frac{16}{8}$ _____

   _____

   (B) $5 + 5 + 3 = 2$ _____

   _____

   (C) $\frac{40}{8} + \frac{40}{8} + \frac{24}{4} = 2$ _____

   _____

   (D) $5 + 5 + 6 = 16$ _____

Meaning of Negative Numbers

Practice Sheet 1

Name: _____

Write the temperature reading indicated on the thermometer and then write a signed number to represent the temperature reading.

Example

Temperature reading:

1 degree below zero

Signed number: ___*-1*___

1.

Temperature reading:

_____

Signed number: _____

2.

Temperature reading:

_____

Signed number: _____

3.

Temperature reading:

_____

Signed number: _____

4.

Temperature reading:

_____

Signed number: _____

5.

Temperature reading:

_____

Signed number: _____

6.

Temperature reading:

_____

Signed number: _____

Meaning of Negative Numbers

Practice Sheet 2

Name: _____

Fill in the thermometer so that it shows the indicated temperature reading and then write a signed number to represent the temperature reading.

Example

Temperature reading:

6 degrees above 0

Signed number: ____+6____

1.

Temperature reading:

5 degrees above 0

Signed number: _____

2.

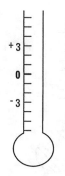

Temperature reading:

4 degrees above ⁻3

Signed number: _____

3.

Temperature reading:

8 degrees below ⁺4

Signed number: _____

4.

Temperature reading:

3 degrees below ⁻1

Signed number: _____

5.

Temperature reading:

1 degree above ⁻4

Signed number: _____

6.

Temperature reading:

7 degrees above ⁻2

Signed number: _____

Meaning of Negative Numbers

Practice Sheet 3

Name: _____   ALGE|B|R|I|D|G|E|™

Describe the position of the arrow on the scale and then write a signed number to represent the scale reading.

Example

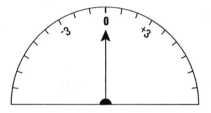

Scale reading: *2 units to the right of*

zero

Signed number: *+2*

1.

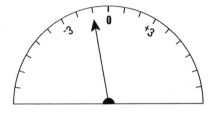

Scale reading: _____

_____

Signed number: _____

2.

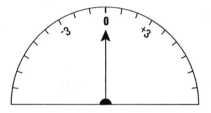

Scale reading: _____

_____

Signed number: _____

3.

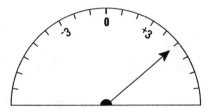

Scale reading: _____

_____

Signed number: _____

4.

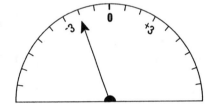

Scale reading: _____

_____

Signed number: _____

5.

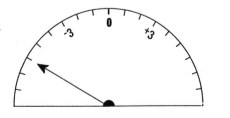

Scale reading: _____

_____

Signed number: _____

6.

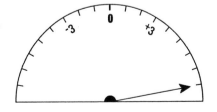

Scale reading: _____

_____

Signed number: _____

Meaning of Negative Numbers

Practice Sheet 4

Name: _____

Draw an arrow on the scale so that it shows the indicated scale reading and then write a signed number to represent the scale reading.

Example

Scale reading: 3 units to the left of ⁺1

Signed number: ‾2 _____

1.

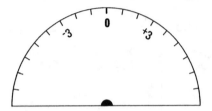

Scale reading: 4 units to the right of ⁺3

Signed number: _____

2.

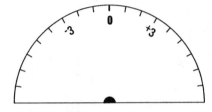

Scale reading: 7 units to the left of ⁺4

Signed number: _____

3.

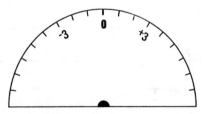

Scale reading: 2 units to the left of ‾3

Signed number: _____

4.

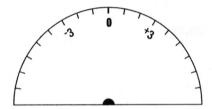

Scale reading: 6 units to the right of ‾3

Signed number: _____

5.

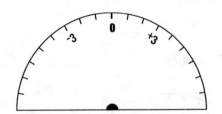

Scale reading: 2 units to the left of ⁺5

Signed number: _____

6.

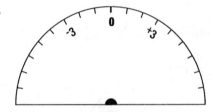

Scale reading: 1 unit to the right of ‾8

Signed number: _____

Meaning of Negative Numbers

Practice Sheet 5

Name: _____   A L G E |B|R|I|D|G|E|™

a)

Temperature reading:

_____

_____

Signed number:

_____

Temperature reading:

_____

_____

Signed number:

_____

b)

Temperature reading:

_____

_____

Signed number:

_____

Temperature reading:

_____

_____

Signed number:

_____

c)

Temperature reading:

_____

_____

Signed number:

_____

Temperature reading:

_____

_____

Signed number:

_____

d)

Temperature reading:

_____

_____

Signed number:

_____

Temperature reading:

_____

_____

Signed number:

_____

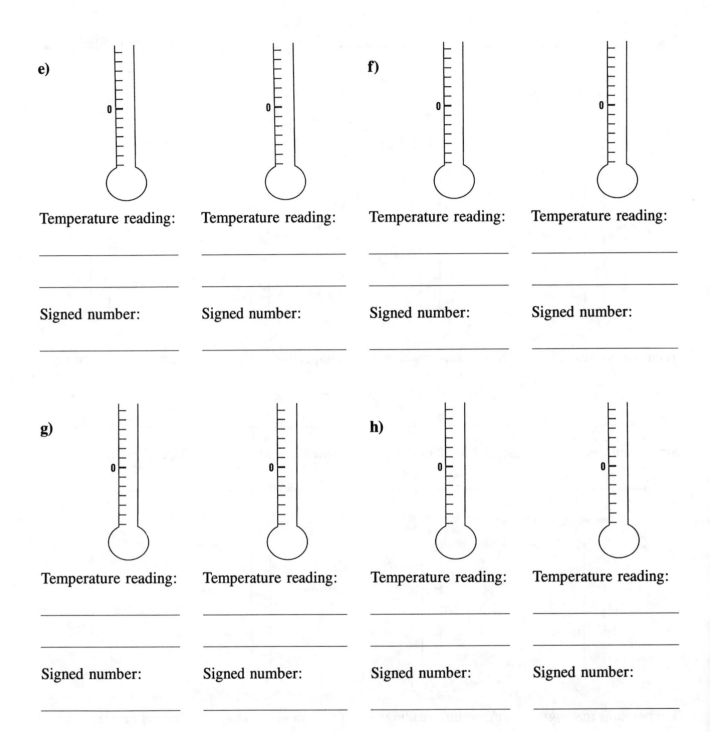

e)

Temperature reading:

_____

_____

Signed number:

_____

Temperature reading:

_____

_____

Signed number:

_____

f)

Temperature reading:

_____

_____

Signed number:

_____

Temperature reading:

_____

_____

Signed number:

_____

g)

Temperature reading:

_____

_____

Signed number:

_____

Temperature reading:

_____

_____

Signed number:

_____

h)

Temperature reading:

_____

_____

Signed number:

_____

Temperature reading:

_____

_____

Signed number:

_____

Meaning of Negative Numbers

Practice Sheet 6

Name: _____

Label the thermometer with three temperatures. Write the temperature reading and then write a signed number to represent the temperature reading. Each reading should be a different number.

Example

0
- 3
- 6

Temperature reading:

3 degrees below zero

Signed number: ⁻3

1.

Temperature reading:

_____

Signed number: _____

2.

Temperature reading:

_____

Signed number: _____

3.

Temperature reading:

_____

Signed number: _____

4.

Temperature reading:

_____

Signed number: _____

5.

Temperature reading:

_____

Signed number: _____

6.

Temperature reading:

_____

Signed number: _____

Meaning of Negative Numbers

Practice Sheet 7

Name: _____   A L G E B R I D G E™

Write a signed number to represent the given temperature reading. Then label the thermometer with three temperatures so that the scale you create allows you to fill in the thermometer so that it indicates the given temperature reading.

Example:

Temperature reading:

4 degrees above −7

Signed number: _−3_

1.

Temperature reading:

7 degrees below 0

Signed number: _____

2.

Temperature reading:

0 degrees

Signed number: _____

3.

Temperature reading:

2 degrees below ⁺5

Signed number: _____

4.

Temperature reading:

3 degrees above ⁺6

Signed number: _____

5.

Temperature reading:

5 degrees

Signed number: _____

6.

Temperature reading:

8 degrees below ⁺4

Signed number: _____

Meaning of Negative Numbers

Practice Sheet 8 — Game Pieces

Meaning of Negative Numbers

Practice Sheet 9 — Scoring Sheet for Card Game

Name: _____

	Number of $+$ Cards Pulled	Number of $-$ Cards Pulled	Score	Name of Winner
Game 1				
Name of player 1				
_____	_____	_____	_____	
Name of player 2				
_____	_____	_____	_____	_____
Game 2				
Name of player 1				
_____	_____	_____	_____	
Name of player 2				
_____	_____	_____	_____	_____
Game 3				
Name of player 1				
_____	_____	_____	_____	
Name of player 2				
_____	_____	_____	_____	_____
Game 4				
Name of player 1				
_____	_____	_____	_____	
Name of player 2				
_____	_____	_____	_____	_____

Meaning of Negative Numbers

Practice Sheet 10

Name: _____

Label the dial with three signed numbers. Write the scale reading indicated by the arrow and then write a signed number that represents the scale reading. Each reading should be a different number.

Example

Scale reading: _One-half unit to the right_

of zero _____

Signed number: $+\frac{1}{2}$

1.

Scale reading: _____

_____

Signed number: _____

2.

Scale reading: _____

_____

Signed number: _____

3.

Scale reading: _____

_____

Signed number: _____

4.

Scale reading: _____

_____

Signed number: _____

5.

Scale reading: _____

_____

Signed number: _____

6.

Scale reading: _____

_____

Signed number: _____

Meaning of Negative Numbers

Practice Sheet 11

Name: _____

Write a signed number to represent the scale reading. Label the dial with three signed numbers so that the scale you create allows you to draw an arrow on the dial that points to the given scale reading.

Example

Scale reading: $\frac{2}{3}$ of a unit to the left of $^-1$

Signed number: ___$-1\frac{2}{3}$___

1.

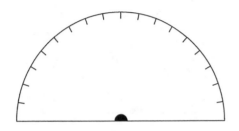

Scale reading: 1 unit to the right of $^+\frac{1}{2}$

Signed number: _____

2.

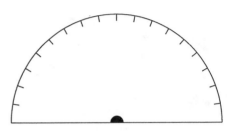

Scale reading: $\frac{1}{2}$ unit to the right of $^-3$

Signed number: _____

3.

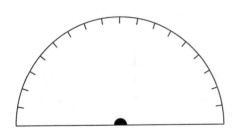

Scale reading: $3\frac{1}{2}$ units to the left of 0

Signed number: _____

4.

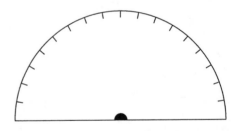

Scale reading: $\frac{3}{4}$ unit to the left of $^-1$

Signed number: _____

5.

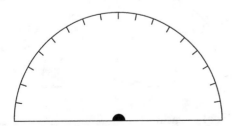

Scale reading: $4\frac{2}{3}$ units to the right of $^+\frac{2}{3}$

Signed number: _____

6.

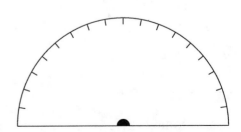

Scale reading: $4\frac{1}{2}$ units to the left of $^+2$

Signed number: _____

Meaning of Negative Numbers

Practice Sheet 12

Name: _____

Label the dial with three signed numbers. Write the scale reading indicated by the arrow and then write a signed number that represents the scale reading. Each reading should be a different number.

Example

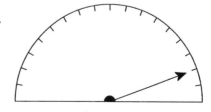

Scale reading: *One-half unit to the right of zero*

Signed number: *+.5*

1.

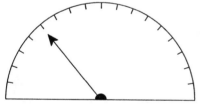

Scale reading: _____

_____

Signed number: _____

2.

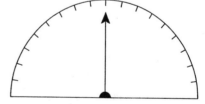

Scale reading: _____

_____

Signed number: _____

3.

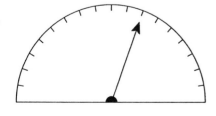

Scale reading: _____

_____

Signed number: _____

4.

Scale reading: _____

_____

Signed number: _____

5.

Scale reading: _____

_____

Signed number: _____

6.

Scale reading: _____

_____

Signed number: _____

Meaning of Negative Numbers

Practice Sheet 13

Name: _____

Write a signed number to represent the scale reading. Label the dial with three signed numbers so that the scale you create allows you to draw an arrow on the dial that points to the given scale reading.

Example

Scale reading: 1 unit to the right of 0.25

Signed number: _+1.25_____

1.

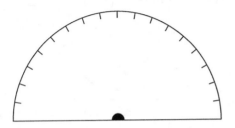

Scale reading: 1 unit to the right of 0.5

Signed number: _____

2.

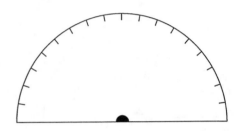

Scale reading: 0.5 unit to the right of ⁻3

Signed number: _____

3.

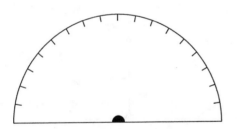

Scale reading: 1.75 units to the left of 0

Signed number: _____

4.

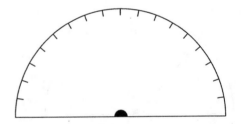

Scale reading: 0.5 unit to the left of ⁻1

Signed number: _____

5.

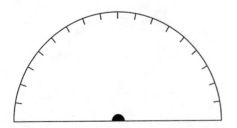

Scale reading: 1.25 units to the right of 0.25

Signed number: _____

6.

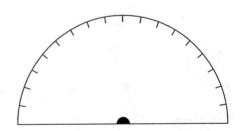

Scale reading: 4.5 units to the left of ⁺2

Signed number: _____

Pattern Recognition and Proportional Reasoning

Practice Sheet 1

Name: _____

Answer each of the following questions.

Questions 1–2 refer to the following figure.

```
        2              Row 1
      3 − 3            Row 2
    2 − 2 + 2          Row 3
  3 − 3 + 3 − 3        Row 4
2 − 2 + 2 − 2 + 2      Row 5
```

1. If the pattern in the triangle is continued as shown and the indicated operations of addition and subtraction are performed, what will be the sum of the numbers in Row 34?

    (A)   0
    (B)   2
    (C)   3
    (D)   102

2. In the triangle, what is the sum of the numbers in Row 63?

    (A)   0
    (B)   2
    (C)   3
    (D)   102

3.

```
            2                Row 1
       2    4    2           Row 2
    2   8    8    2          Row 3
  2  16   64   16   2        Row 4
2  □  1024  1024  □  2       Row 5
```

In the triangle shown above, what number should replace the □ if the pattern remains the same?

    (A)   16
    (B)   32
    (C)   64
    (D)   128

4. Draw a figure that could come next in the sequence below.

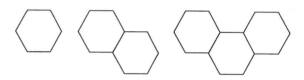

_____

5. a)   1, 2, 3, 4

In the list of numbers above, what number is added to each number in the list to get the next number?

    (A)   0
    (B)   1
    (C)   2

b)   1, 2, 3, 4, 5

In the list of numbers above, what number is added to each odd number in the list to get the next larger odd number?

    (A)   0
    (B)   1
    (C)   2

6.   1, 3, 2, 4, 5, 7, 6, 8, 9,...

The numbers above form a pattern. What is the next number in the pattern?

    (A)   10
    (B)   11
    (C)   12
    (D)   13

Pattern Recognition and Proportional Reasoning

Practice Sheet 2

Name: _____

For questions 1–3, divide each group of 12 marbles into two smaller groups that are in the given ratio.

Example

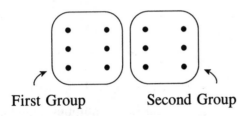

Ratio is 6 to 6 or $\frac{6}{6}$ or $\frac{1}{1}$.

Answer

First Group   Second Group

1. Ratio is 1 to 3

2. Ratio is 2 : 1

3. Ratio is $\frac{1}{5}$

4. Can a group of 12 marbles be divided into two smaller groups that are in the ratio of

   1 to 4? _____   2 to 3? _____

Pattern Recognition and Proportional Reasoning

Practice Sheet 3

Name: _____ ALGE|B|R|I|D|G|E|™

Answer each of the following questions.

1. Harry owns four times as many cats as dogs. Consider each of the following. Could it be the number of cats and dogs that Harry owns?

			Cats	Dogs
yes	no	**a)**	8	2
yes	no	**b)**	2	8
yes	no	**c)**	3	6
yes	no	**d)**	10	6
yes	no	**e)**	20	5

2. In a certain type of salad dressing, there are 2 tablespoons of oil for each tablespoon of vinegar. What is the ratio of the number of tablespoons of oil to the number of tablespoons of vinegar?

   (A) 2 to 1
   (B) 1 to 2
   (C) 1 to 3
   (D) 2 to 3

3. The figure shows 100 small squares. What is the ratio of the number of shaded small squares to the total number of small squares?

   (A) $\dfrac{3}{10}$

   (B) $\dfrac{3}{30}$

   (C) $\dfrac{3}{97}$

   (D) $\dfrac{3}{100}$

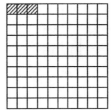

4. In the puzzle, what is the ratio of the number of triangular pieces to the total number of pieces?

   (A) 5 to 7
   (B) 5 to 2
   (C) 7 to 5
   (D) 2 to 5

5. Judy earned $63 in 6 weeks working part-time. If she earned the same amount of money each week, how much did she earn in 4 weeks?

   (A) $10
   (B) $42
   (C) $252
   (D) $378

Pattern Recognition and Proportional Reasoning

Practice Sheet 4

Name: _____   A L G E |B|R|I|D|G|E™

For each question, shade the correct number of blocks in the square on the right so that the ratio of the number of shaded blocks to the total number of blocks is the same for both squares. Then complete the ratio on the right by filling in the top number.

1.

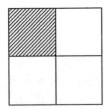 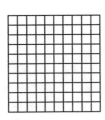

$$\frac{1}{4} \quad = \quad \frac{}{100}$$

2.

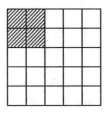 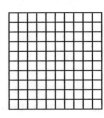

$$\frac{4}{25} \quad = \quad \frac{}{100}$$

3.

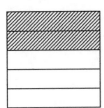 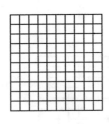

$$\frac{2}{5} \quad = \quad \frac{}{100}$$

4.

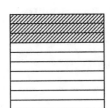 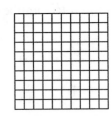

$$\frac{3}{10} \quad = \quad \frac{}{100}$$

5.

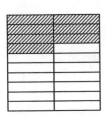 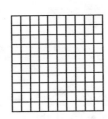

$$\frac{7}{20} \quad = \quad \frac{}{100}$$

6.

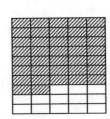 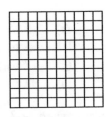

$$\frac{37}{50} \quad = \quad \frac{}{100}$$

Pattern Recognition and Proportional Reasoning

Practice Sheet 5

Name: _____ A L G E |B|R|I|D|G|E|™

Answer each of the following questions.

1.

Column A	Column B
1	5
2	7
3	9
5	13

a) What is the rule that relates each number in column A to the corresponding number in column B in the table above?

_____

b) Write the rule in symbols, letting □ stand for a number in column A and △ stand for the corresponding number in column B.

_____

c) If 4 is placed in column A, what corresponding number should be placed in column B?

_____

d) Use your rule to complete the table below

Column A	Column B
9	_____
13	_____

2.

Column A	Column B
1	5
2	8
3	11
4	14
5	17
6	20

a) What is the rule that relates each number in column A to the corresponding number in column B in the table above?

_____

b) Using a calculator and your rule, extend the table as directed by your teacher. Write your answers in the space below.

Pattern Recognition and Proportional Reasoning

Practice Sheet 6

Name: _____

Answer each of the following questions.

• •
• •   • •
• •   • •   • •

Shape 1   Shape 2   Shape 3   Shape 4

1. Draw shape 4 for the pattern in the space provided above.

2. Write a rule that describes how to find the total number of dots for each new shape in the pattern.

   Rule: _____

3. Complete the table below that relates the number of each shape in the pattern to the number of dots in that shape.

Number of Shape	Number of Dots in Shape
1	2
2	_____
3	_____
4	_____
5	_____
6	_____

4. How many dots will form the 10th shape? _____

   How many dots will form the 43rd shape? _____

5. If □ stands for the number of a shape in the pattern and △ stands for the total number of dots in that shape, write a rule that uses □ and △.

   Rule: _____

6. What is the rule for finding the total number of dots for each new shape in the pattern below?

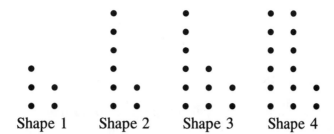

Shape 1   Shape 2   Shape 3   Shape 4

Rule: _____

_____

7. In the space below, draw the first 4 shapes of a dot pattern in which the total number of dots for each shape can be found by multiplying the shape number by one number and adding another number. Describe the pattern to another student in your class so that he or she can draw it. The result should be the same as your pattern.

Pattern Recognition and Proportional Reasoning

Practice Sheet 7

Name: _____   A L G E |B|R|I|D|G|E|™

Answer each of the following questions.

You will need a sheet of graph paper and a pair of scissors to do this activity. For this activity, 1 unit will be the same as the length of a side of 1 block on the graph paper.

1. Cut out of the graph paper all rectangles shown in the pattern below.

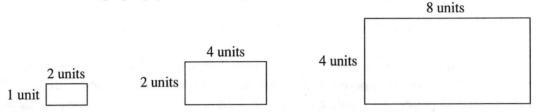

2. Cut out the fourth rectangle (not shown) in the pattern. Describe it.

_____

One way to determine whether a rectangle belongs in the pattern above is to investigate whether a number of the smaller rectangles in the pattern will fit completely, without overlapping, inside the rectangle under consideration.

For example, the figure below shows that four rectangles, each measuring 1 unit by 2 units, will fit completely inside the 2 unit by 4 unit rectangle, or sixteen rectangles, each measuring 1 unit by 2 units, will fit completely inside the 4 unit by 8 unit rectangle.

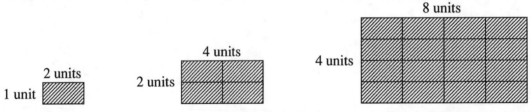

How many rectangles, each measuring 2 units by 4 units, will fit completely inside the 4 unit by 8 unit rectangle?

_____

Now cut out a rectangle that measures 3 units by 12 units. Use the method of fitting smaller rectangles into the 3 unit by 12 unit rectangle to determine whether this rectangle belongs in the pattern above.

3. Cut out two more rectangles using widths and lengths of your choice. Are they a part of the pattern above?

_____

4. Look at the lengths of the longer and shorter sides for each rectangle that is in the pattern above. How can you determine whether any new rectangle that is drawn is a part of the pattern?

_____

_____

5. Complete the following table for the pattern above.

Width	Length
_____	_____
_____	1
1	2
2	_____
4	8
8	_____
16	_____
_____	_____

6. Describe another method, other than the one used above, to determine whether any rectangle is a part of the pattern.

_____

_____

_____

Pattern Recognition and Proportional Reasoning

Practice Sheet 8

Name: _____

Answer each question.

Questions 1–3 refer to the following situation.

Roller skates are rented for a fee of $2.00 plus $.50 per hour for every hour they are used.

1. Write the rule in words that could be used to find the cost of renting a pair of roller skates for 4 hours.

_____

2. If □ stands for the number of hours and △ stands for the total cost, write the rule in symbols that could be used to find the cost of renting a pair of roller skates.

_____

3. For how many hours can a pair of skates be rented for $5.00?

_____

Questions 4–6 refer to the following situation.

Do-It-Right Print Shop told the ticket committee for the local school play that the cost of printing tickets would be $.20 per ticket plus a base charge of $50.00 for typesetting.

4. Write a rule in words that could be used to find the cost of printing 500 tickets.

_____

5. If □ stands for the number of tickets and △ stands for the total cost of printing tickets, write the rule in symbols that could be used to find the cost of printing tickets.

_____

6. How many tickets can be printed for $175.00?

_____

Pattern Recognition and Proportional Reasoning

Practice Sheet 9

Name: _____

Answer each of the following questions.

1.

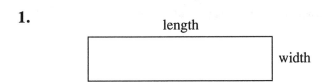

length

width

The area of a rectangle is found by multiplying length times width. What happens to the area of a rectangle when its length and width are both tripled?

_____

2.

Column A	Column B
5	15
12	36
4	12
9	27

a) Look at the table above. What rule can be applied to each number in column A to get the corresponding number in column B?

_____

_____

b) Fill in the missing numbers in the table.

Column A	Column B
5	15
12	36
4	12
9	27
2	_____
35	_____
8	_____

c) Place a number in column A that is different from those already shown there. If the rule determined in part a is used, what is the corresponding number in column B?

_____

d) Now increase the number that you placed in column A in part c by 2. What happens to the corresponding number in column B? Does it increase or decrease? By how much?

_____

e) Suppose you increase the number that you placed in column A by 28. Determine without calculating what will happen to the corresponding number in column B.

_____

3.

Column A	Column B
1	4
2	8
3	12
4	16
5	20

In the table above, each number in column B is 4 times the corresponding number in column A. If 1 is added to each number in column A, what number should be added to each number in column B so that each number in column B is still 4 times the corresponding number in column A?

_____

Constructing Numerical Equations

Practice Sheet 1

Name: _____

Use the information given to label each diagram.

1. Sandra is half as old as her mother. Which diagram could represent Sandra's age? Which diagram could represent her mother's age?

    _____'s age   _____'s age

2. Number *A* is 9 less than number *B*. Label each diagram with the number it could represent.

    _____   _____

3. Darlene has 3 times as much money as Sylvia. Which diagram could represent the amount of money that Sylvia has? Which diagram could represent the amount of money that Darlene has?

    Amount that _____ has   Amount that _____ has

4. Last week, Neil delivered 12 more newspapers than Tom delivered. Which diagram could represent the number of newspapers delivered by Neil? Which diagram could represent the number of newspapers delivered by Tom?

    Number delivered by _____   Number delivered by _____

Constructing Numerical Equations

Practice Sheet 2

Name: _____

Draw diagrams based on the information given.

1. Yesterday's high temperature was 8 degrees more than today's high temperature. Draw diagrams to represent the high temperature for each of the two days.

2. Mona has half as much money as Mike. Draw diagrams to represent the amount of money that each person has.

3. Mario is 15 years older than Bill. Draw diagrams to represent the age of each person.

4. Two numbers differ by 16. Draw diagrams to represent each number.

5. There are 5 more girls than boys who play brass instruments in the band. Draw diagrams to represent the number of each who play brass instruments.

Constructing Numerical Equations

Practice Sheet 3

Name: _____

Label each diagram so that it represents the situation in the problem.

1. In a road rally, the total distance traveled by car *A* was 4 times as much as the total distance traveled by car *B*.

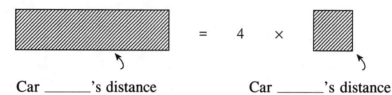

Car _____'s distance   Car _____'s distance

2. Last month's total rainfall in a certain town was less than this month's total rainfall in the same town.

_____ month's rainfall   _____ month's rainfall

3. On a math quiz, Carmella's score was 5 points more than Rob's score.

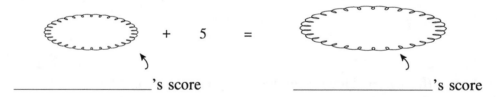

_____'s score   _____'s score

4. James bought two records and a cassette tape. He spent $15.98 for the three items.

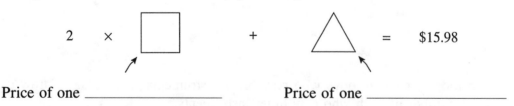

Price of one _____   Price of one _____

5. If 8 more than a number is divided by the number, the quotient is 3.

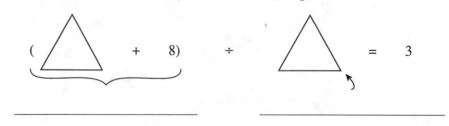

_____   _____

Constructing Numerical Equations

Practice Sheet 4

Name: _____

Draw and label diagrams that represent the situation in each problem.

1. The sum of two numbers is 54.

2. Leslie has 3 times as many baseball cards in her collection as Tonya has in her collection.

3. The number of people attending the Saturday evening performance of the school play was equal to 83 more than the number of people attending the Friday evening performance.

4. Chuck is half as tall as his older brother Mark.

5. The number of miles between Lan's house and school is 7 less than the number of miles between Cara's house and school.

Constructing Numerical Equations

Practice Sheet 5

Name: _____

Write a problem that could be represented by each diagram.

1.

$$\triangle \; + \; 12 \; = \; \square$$

Number of boys   Number of girls

_____

_____

_____

2.

$$\bigcirc \; = \; 9 \; \times \; \bigcirc$$

Number of red marbles   Number of blue marbles

_____

_____

_____

3.

$$\square \; - \; \square \; = \; 70$$

A number   One half the number

_____

_____

_____

Constructing Numerical Equations

Practice Sheet 6

Name: _____ ALGE|B|R|I|D|G|E|™

Complete each table based on the information given.

1. The second number is 4 more than the first number.

First Number	Second Number
1	5
3	_____
_____	14
_____	_____
_____	_____

2. Alice is 5 years younger than Ron.

Alice's Age	Ron's Age
20	25
14	_____
_____	14

3. Lucia earned twice as much last week as she did this week.

Amount Earned Last Week	Amount Earned This Week
$400	$200
$124	_____
_____	$390
_____	_____
_____	_____

4. The ratio of two debts is 5 to 3.

Amount of First Debt	Amount of Second Debt
$5	$3
$30	$18
$15	_____
$80	_____
_____	$27
_____	_____
_____	_____

Constructing Numerical Equations

Practice Sheet 7

Name: _____

Make a table based on the information given in each problem. Be sure to label each column in the table.

1. Number *A* is 5 times number *B*.

_____   _____

    _____   _____

    _____   _____

    _____   _____

    _____   _____

    _____   _____

2. Maria is 9 years older than Keith.

_____   _____

    _____   _____

    _____   _____

    _____   _____

    _____   _____

    _____   _____

3. Akia has one third as many records as Linda.

_____   _____

    _____   _____

    _____   _____

    _____   _____

    _____   _____

    _____   _____

4. The number of students in a certain algebra class is at least 10 more than the number of students in a certain geometry class.

_____   _____

    _____   _____

    _____   _____

    _____   _____

    _____   _____

    _____   _____

Constructing Numerical Equations

Practice Sheet 8

Name:_____

Complete each table based on the information given.

1. The difference between two numbers is 14.

First Number	−	Second Number	=	14
15	−	1	=	14
29	−	_____	=	14
82	−	_____	=	14
_____	−	65	=	14
_____	−	140	=	14
_____	−	_____	=	14
_____	−	_____	=	14

2. Lydia spent $13 more than Paul.

Amount Lydia Spent	=	13	+	Amount Paul Spent
32	=	13	+	19
_____	=	13	+	5
_____	=	13	+	58
90	=	13	+	_____
211	=	13	+	_____
_____	=	13	+	_____
_____	=	13	+	_____

3. In a certain school, 5 times as many students buy lunch in the cafeteria as bring lunch from home.

5(Number Who Bring from Home)	=	Number Who Buy in Cafeteria
5(17)	=	85
5(8)	=	_____
5(_____)	=	135
5(_____)	=	80
5(43)	=	_____
5(_____)	=	_____
5(_____)	=	_____

Constructing Numerical Equations

Practice Sheet 9

Name: _____

Make a table based on the information given in each problem. Remember to label each column of the table.

1. The time it took Lars to run a mile was 4 seconds less than the time it took Jan to run a mile.

   _____   _____

   _____   _____

   _____   _____

   _____   _____

   _____   _____

   _____   _____

2. If the number of people who are current members of the Neighborhood Sports Center is divided by 8, the result will be the same as the number of original members of the Neighborhood Sports Center.

   _____   _____

   _____   _____

   _____   _____

   _____   _____

   _____   _____

   _____   _____

3. The price of an ice-cream sandwich is one-third the price of an ice-cream sundae.

   _____   _____

   _____   _____

   _____   _____

   _____   _____

   _____   _____

4. Mrs. Mason is 23 years older than her son Tim.

   _____   _____

   _____   _____

   _____   _____

   _____   _____

   _____   _____

Constructing Numerical Equations

Practice Sheet 10

Name: _____

Write a statement that could be represented by each table.

1.

Number X	$=$	12(Number Y)
48	$=$	12(4)
156	$=$	12(13)
60	$=$	12(5)
96	$=$	12(8)

_____

_____

_____

2.

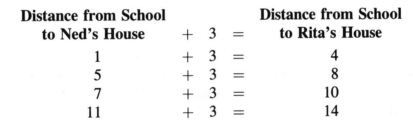

Distance from School to Ned's House	$+$	3	$=$	Distance from School to Rita's House
1	$+$	3	$=$	4
5	$+$	3	$=$	8
7	$+$	3	$=$	10
11	$+$	3	$=$	14

_____

_____

_____

3.

Number of Hamburgers Sold	$-$	Number of Hot Dogs Sold	$=$	18
23	$-$	5	$=$	18
46	$-$	28	$=$	18
109	$-$	91	$=$	18
325	$-$	307	$=$	18

_____

_____

_____

Constructing Numerical Equations

Practice Sheet 11

Name: _____   A L G E |B|R|I|D|G|E|™

For each of the following, determine what additional information is required to solve the problem and write that information in the space provided below the problem.

1. Miguel wants to buy a stereo system. He has saved $132. How much more money does he need to buy the stereo system?

   _____

   _____

2. For each of 5 consecutive days last week, Connie ran 1 mile farther than she did on the preceding day. What was the total distance, in miles, that Connie ran last week?

   _____

   _____

3. In a basketball game between Central School and Main Street School, the team from Central School scored twice as many points as the team from Main Street School. How many points were scored by the Central School team?

   _____

   _____

4. On a map a length of 1 inch represents a distance of 10 miles. What is the distance between the cities of Fernville and Oceantown?

   _____

   _____

5. Last Friday Nicki's Pizza Shop sold 40 medium pizzas and 30 large pizzas. The total profit earned on these 70 pizzas was $380. If Nicki's Pizza Shop earns more profit on a large pizza than on a medium pizza, how much profit is earned on one large pizza?

   _____

   _____

Constructing Numerical Equations

Practice Sheet 12

Name: _____ ALGE|B|R|I|D|G|E|™

For each of the following, cross out any information that is **not** required to solve the problem. (Do not solve the problem!)

1. The student council is taking a field trip to an amusement park. Admission tickets are $8 for students and $10 for adults. There are 53 students and 5 chaperones going on the trip in 2 buses. What is the total amount required to purchase tickets for this group?

2. The length and width of a rectangle are 9 and 8, and its perimeter is 34. If the length and width are doubled, what is the area of the new rectangle that is formed?

3. A certain automobile can travel 21 miles on a gallon of gasoline. How many stops to fill the fuel tank will be required on a trip of 1,350 miles if gasoline costs $1.15 per gallon and the tank holds 20 gallons?

4. Town A gets an average of 13 inches of rainfall per year, and Town B gets an average of 44 inches of rainfall per year. There are twice as many cloudy days in Town B as in Town A. What is the difference in average rainfall per year between Town A and Town B?

5. In a basketball game, 25% of the points scored by one team were scored by their star player. If that team won the game by 3 points and the losing team scored a total of 97 points, how many points were scored by the star player?

Attacking Word Problems Successfully

Practice Sheet 1

Name: _____

Read the following problems. Determine which information that is given is not needed to find the answer. Then cross out that information.

1. Julie has $14.50 in her pocket. She wants to buy a record that costs $6.50. Three weeks ago, she bought a book for $7.95. How much change should she receive from the purchase of the record?

2. The gas mileage of Mark's moped is 60 miles per gallon. The capacity of the gas tank is 1.5 gallons. How many gallons of gas will Mark need to travel 15 miles?

3. In a class of 15 students, there are 9 boys and 6 girls. The students' scores on a recent mathematics test were 87, 92, 80, 95, 72, 82, 65, 63, 74, 80, 88, 77, 98, 84, 71. What was the class average for this test?

4. Anita bought 2 necklaces on Saturday at a special sale. Earrings were on sale for $8.95 a pair. If Anita received $12.98 in change from the $50 she gave the sales clerk, how much did the 2 necklaces cost?

5. Write a problem of your own that includes information that is needed and information that is not needed to find the answer. Exchange papers with a classmate. For the problem that your classmate wrote, determine which information is not needed to find the answer. Then cross out that information.

Attacking Word Problems Successfully

Practice Sheet 2

Name: _____

For each exercise below, show two different ways to divide the area of the figure into rectangles by drawing lines inside each figure.

1.

2.

3.

4.

Attacking Word Problems Successfully

Practice Sheet 3

Name: _____

This activity requires you to work with another student in your class. You and your partner will each be given a design to describe to the other. Without showing your design to your partner, sketch your design on this paper and write a description in the space provided. Now, one of you will describe while the other attempts to draw the design. The person describing must use words only—no pointing or other hand gestures! The person who is drawing must not ask questions. When you finish describing and drawing, discuss what was helpful or confusing about the description. Now switch roles and describe and draw the other design.

Sketch your design here.

Description: _____

_____

_____

_____

_____

_____

_____

Attacking Word Problems Successfully

Practice Sheet 3A—Designs for use with practice sheet 3

To the teacher: After cutting out the designs below, you may wish to distribute one design to each student or group. Alternatively, the cut-out designs could be placed in a container and each student could randomly select a design. You may need to remind students not to show their designs to their partners until the appropriate time in the activity.

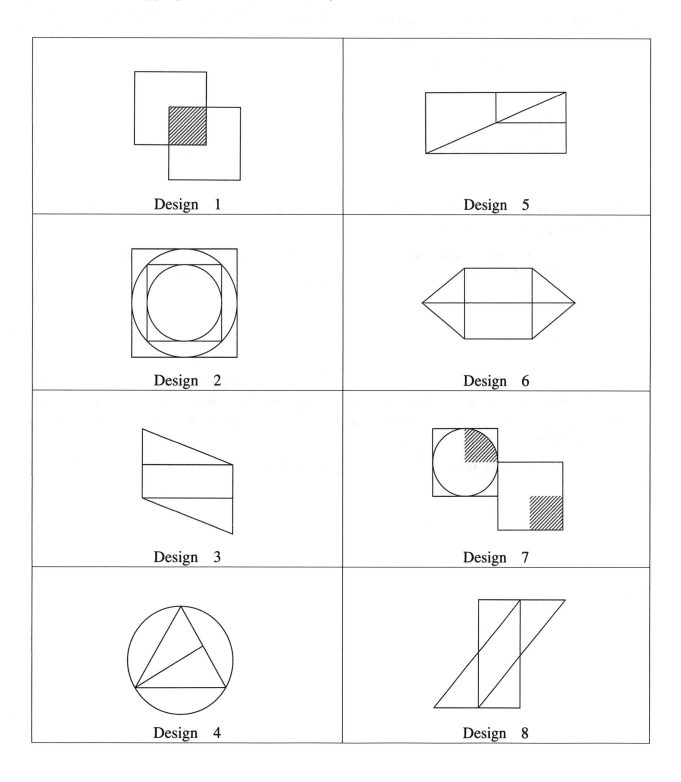

Design   1

Design   5

Design   2

Design   6

Design   3

Design   7

Design   4

Design   8

Attacking Word Problems Successfully

Practice Sheet 4

Name: _____ A L G E | B | R | I | D | G | E |™

In each of the following problems, complete the table to help you determine the answer.

1. If one of three consecutive whole numbers is 53, what could be the sum of the three numbers?

Three Consecutive Whole Numbers	Sum

2. The sum of two numbers is equal to 50. If one of the numbers is divisible by both 3 and 4, what could be the other number?

Numbers Less Than 50 That Are Divisible by 3 and 4	Number Which When Added Gives a Sum of 50

3. A cash drawer contains only $1 and $5 bills. If there is at least one of each kind of bill in the drawer and the total number of bills is 9, what is the least amount of money in $1 and $5 bills in the drawer?

Number of $1 Bills	Number of $5 Bills	Total Value

Attacking Word Problems Successfully
Practice Sheet 5

Name: _____

Solve the following problems.

1. Alice, Brent, Candy, and Dale each participate in a different sport. Their sports are track, gymnastics, swimming, and basketball. Candy is the sister of the person whose sport is gymnastics. Water is required for Alice to participate in her sport. While engaging in his sport, Dale once took part in a 400-meter race. Brent won a trophy in his sport last month. Determine each student's sport.

2. Zeke the grocer follows several rules when he displays his vegetables. For example, since leaf lettuce must be sprayed periodically to keep it looking fresh, it must be placed in the first bin on the left. Otherwise, the hose won't reach. Tomatoes are always placed next to the lettuce, but onions are never next to potatoes. Zeke puts carrots in the right-most bin because they are least likely to roll off. Also, for overall appearance, Zeke never places two green vegetables next to each other.

   If Zeke has 7 bins, give one possible arrangement for onions, potatoes, lettuce, carrots, green beans, broccoli, and tomatoes.

Attacking Word Problems Successfully

Practice Sheet 6

Name: _____

Solve the following problem.[1]

A hiker, carrying no equipment, got lost during a summer hike and fell into a circular pit that was about 30 feet across and 20 feet deep. All the walls of the pit were hard, very smooth stone, and were straight up and down, with no handholds or footholds for climbing. The bottom of the pit was also hard stone. A stream of water flowed over the edge of the pit and down one wall. Behind the waterfall, the wall was smooth and straight up and down. After reaching the bottom, the water disappeared into a small hole in the floor of the pit. The pit was completely empty except for three things:

1. In the exact middle of the pit, a tree grew straight up.
2. Near the tree was a loose flat rock.
3. Two 6-foot boards were also near the tree.

How did the hiker escape from the pit?

[1] *The Arithmetic Teacher* 31.4 (March 1984): 41-43.

Attacking Word Problems Successfully

Practice Sheet 7

Name: _____   A L G E |B|R|I|D|G|E| ™

Survey a group of people to determine their favorite musical group, favorite type of food, or other piece of information of interest to you. Record the plans for, and the results of, your survey below. Use additional paper if necessary.

Topic: _____

Description of people to be surveyed: _____

_____

Questions to be asked: _____

_____

_____

_____

Summary of results: _____

_____

_____

_____

Conclusions: _____

_____

_____

Concept of Variable

Practice Sheet 1

Name: _____

Write each number in three different ways. After you have completed this exercise, compare your answers with those of another student in the class.

Example   1,988

Possible Answers $2,000 - 12$

$1,388 + 600$

$5,964 \div 3$

1.   14   _____

_____

_____

2.   0   _____

_____

_____

3.   295   _____

_____

_____

4.   888   _____

_____

_____

5.   1,776   _____

_____

_____

6.   10,000   _____

_____

_____

Concept of Variable

Practice Sheet 2

Name: _____

Using only the digits in the number 1,989 and the operation symbols, form each of the numbers from 1 to 100. The digits must be kept in their original order and no other digits may be used. For example, $27 = 1 + 9 + 8 + 9$ and $36 = 19 + 8 + 9$.

1 = _____	26 = _____	51 = _____	76 = _____
2 = _____	27 = $1 + 9 + 8 + 9$	52 = _____	77 = _____
3 = _____	28 = _____	53 = _____	78 = _____
4 = _____	29 = _____	54 = _____	79 = _____
5 = _____	30 = _____	55 = _____	80 = _____
6 = _____	31 = _____	56 = _____	81 = _____
7 = _____	32 = _____	57 = _____	82 = _____
8 = _____	33 = _____	58 = _____	83 = _____
9 = _____	34 = _____	59 = _____	84 = _____
10 = _____	35 = _____	60 = _____	85 = _____
11 = _____	36 = $19 + 8 + 9$	61 = _____	86 = _____
12 = _____	37 = _____	62 = _____	87 = _____
13 = _____	38 = _____	63 = _____	88 = _____
14 = _____	39 = _____	64 = _____	89 = _____
15 = _____	40 = _____	65 = _____	90 = _____
16 = _____	41 = _____	66 = _____	91 = _____
17 = _____	42 = _____	67 = _____	92 = _____
18 = _____	43 = _____	68 = _____	93 = _____
19 = _____	44 = _____	69 = _____	94 = _____
20 = _____	45 = _____	70 = _____	95 = _____
21 = _____	46 = _____	71 = _____	96 = _____
22 = _____	47 = _____	72 = _____	97 = _____
23 = _____	48 = _____	73 = _____	98 = _____
24 = _____	49 = _____	74 = _____	99 = _____
25 = _____	50 = _____	75 = _____	100 = _____

Concept of Variable

Practice Sheet 3

Name: _____

Circle yes or no for each of the following statements.

1. Consider each of the following. Is the statement true?

    yes   no   **a)** $9 = 9$

    yes   no   **b)** $30 = 9 + 9 + 12$

    yes   no   **c)** $9 + 9 + 12 = 30$

    yes   no   **d)** $9 + 12 = 30$

    yes   no   **e)** $9 + 9 = 30 - 12$

2. Consider each of the following. Is the statement true?

    yes   no   **a)** $\boxed{15} = 18 - 3$

    yes   no   **b)** $\boxed{15} = \boxed{15}$

    yes   no   **c)** $\boxed{15} + 2 + 5 = \boxed{15} + 3 + 4$

    yes   no   **d)** $6 - \boxed{15} = \boxed{15} - 6$

    yes   no   **e)** $\boxed{15} + 4 = \boxed{15} + 5$

    yes   no   **f)** $18 - 3 = \boxed{15}$

3. Consider each of the following. If 9 is placed in the $\square$, is the statement true?

    yes   no   **a)** $15 + \square = 24$

    yes   no   **b)** $39 = 30 + \square$

    yes   no   **c)** $\square + 3 = \square + 4$

    yes   no   **d)** $\square + 5 = 23 - \square$

    yes   no   **e)** $\square + \square = 18$

    yes   no   **f)** $\square = 18 - \square$

4. Consider each of the following. If k is replaced by 5, is the statement true?

    yes   no   **a)** $16 + k = 18$

    yes   no   **b)** $k + 7 = k + 8$

    yes   no   **c)** $42 = 37 + k$

    yes   no   **d)** $k + 15 = 25 - k$

    yes   no   **e)** $k + k = 10$

    yes   no   **f)** $k = 10 - k$

Concept of Variable

Practice Sheet 4

Name: _____   A L G E B R I D G E™

Answer each question in the space provided.

1. If k represents the number of record albums that one student has and all students have the same number of record albums, how would you represent the number of record albums that 3 students have?

   _____

   Now suppose that $k = 14$. How many record albums would the 3 students have?

   _____

2. There are 5 sticks of chewing gum in one pack. If r represents the number of packs of chewing gum, how would you represent the number of sticks of chewing gum in r packs?

   _____

   Another name for $5r$ is $r + r + r + r + r$. If r represents the number of packs of chewing gum, complete the following table.

r	$r + r + r + r + r$	$5 \cdot r$	$5r$
1	_____	_____	_____
2	$2 + 2 + 2 + 2 + 2$	$5 \cdot 2$	10
3	_____	_____	_____
5	_____	_____	_____
12	_____	_____	_____
20	_____	_____	_____

(Note: This table emphasizes that $r + r + r + r + r$, $5 \cdot r$, and $5r$ are equivalent expressions.)

Concept of Variable

Practice Sheet 5

Name: _____

For each inequality, determine the least or greatest whole number that will make the inequality true when that number is placed in the $\square$.

1. The least whole number that will make $\square + 4 > 18$ true is _____ .

2. The least whole number that will make $3 + \square > 7$ true is _____ .

3. The greatest whole number that will make $\square - 23 < 36$ true is _____ .

4. The least whole number that will make $20 - \square < 12$ true is _____ .

5. The least whole number that will make $2 \times \square + 3 > \square + 4$ true is _____ .

6. The greatest whole number that will make $\square + \square + \square < 10$ true is _____ .

7. The least whole number that will make $53 < \square - 10$ true is _____ .

8. The greatest whole number that will make $\square \div 3 < 9$ true is _____ .

Concept of Variable

Practice Sheet 6

Name: _____   A L G E B R I D G E™

Answer each question in the space provided.

1.

Column A	Column B
10	7
12	9
15	12
20	17
23	_____
29	_____
35	_____

2.

Column A	Column B
1	3
2	5
5	11
9	19
16	_____
30	_____
40	_____

a) In the table above, a rule is applied to each number in column A to get the corresponding number in column B. Determine the rule and use it to supply the missing numbers.

b) Describe the rule in words.

_____

_____

c) If *w* stands for any number in column A, write an expression, using *w*, that shows the rule in symbols.

_____

a) In the table above, a rule is applied to each number in column A to get the corresponding number in column B. Determine the rule and use it to supply the missing numbers.

b) Describe the rule in words.

_____

_____

c) If *v* stands for any number in column A, write an expression, using *v*, that shows the rule in symbols.

_____

Concept of Variable

Practice Sheet 7

Name: _____

Answer each question in the space provided.

1. Jeanette has only chickens and horses on her farm. One day, she counted 86 chicken legs in her chicken house; so she know that there were _____ chickens in the chicken house. She also counted 56 horse legs in her barn, which meant that there were _____ horses in the barn. The next day, Jeanette noticed that some of the horses that were in the barn yesterday had moved to the pasture after she had finished counting. She counted _____ horse legs in the pasture, which meant that _____ horses had moved to the pasture.

    The following week, Jeanette sold some of her chickens and horses to the farmer down the road. The total number of chicken and horse legs on Jeanette's farm dropped from 142 before the sale to 130 after the sale. Fill in the following table to show all possible combinations of chickens and horses sold.

Number of Chickens Sold	Number of Horses Sold

Questions 2–5 refer to the following situation:

   Paula has 5 more cassettes in her music collection than Mark has in his music collection. Every month, Paula and Mark each add 1 new cassette to their collections.

2. If Mark starts with 8 cassettes, after 2 months he will have _____ cassettes and Paula will have _____ cassettes.

3. If Mark starts with 8 cassettes, how would you express, using k, the number of cassettes he would have after k months?

    _____

4. If Mark starts with 10 cassettes, how would you express the number of cassettes Paula would have after k months?

    _____

5. If after q months, Paula has $26 + q$ cassettes, how many cassettes did Mark start with q months ago?

    _____

Concept of Variable

Practice Sheet 8

Name: _____

Circle yes or no for each of the following statements.

1.

The perimeter of the square shown above can be found by multiplying 4 times the length of a side. If s stands for the length of a side, then $4s$ is another name for the perimeter of the square.

Consider each of the following. Is the statement true?

yes	no	**a)**	If the length of a side of the square is 5, then $4 \cdot 5$ is another name for $4s$.
yes	no	**b)**	If the length of a side of the square is 5, then $5 + 5 + 5 + 5$ is another name for $4s$.
yes	no	**c)**	If the length of a side of the square is 5, then 20 is another name for $4s$.
yes	no	**d)**	If the length of a side of the square is 5, then 45 is another name for $4s$.
yes	no	**e)**	If the length of a side of the square is 5, then 9 is another name for $4s$.

2. Will is paid 5 cents for each daily edition of a newspaper that he delivers and 8 cents for each Sunday edition of the same newspaper that he delivers. If n stands for the number of daily editions he delivers and r stands for the number of Sunday editions he delivers, then $5n + 8r$ is another name for the total amount of money, in cents, that Will earns delivering newspapers.

Consider each of the following. Is the statement true?

yes	no	**a)**	If Will delivers 30 daily editions and 15 Sunday editions, then $530 + 815$ is another name for $5n + 8r$.
yes	no	**b)**	If Will delivers 30 daily editions and 15 Sunday editions, then $5 \cdot 30 + 8 \cdot 15$ is another name for $5n + 8r$.
yes	no	**c)**	If Will delivers 30 daily editions and 15 Sunday editions, then 120 is another name for $5n + 8r$.
yes	no	**d)**	If Will delivers 30 daily editions and 15 Sunday editions, then 150 is another name for $5n + 8r$.
yes	no	**e)**	If Will delivers 30 daily editions and 15 Sunday editions, then 270 is another name for $5n + 8r$.

Concept of Variable

Practice Sheet 9

Name: _____

Questions 1 and 2 refer to the following situation:

The number of Sunday editions of a newspaper that Will delivers is less than the number of daily editions of the same newspaper that he delivers. If r stands for the number of Sunday editions and n stands for the number of daily editions, then, in symbols, $r < n$.

1. In the table below, consider each pair of numbers. Is $r < n$ for that pair of numbers?

			r	n
yes	no	**a)**	25	28
yes	no	**b)**	19	16
yes	no	**c)**	30	40
yes	no	**d)**	30	20
yes	no	**e)**	45	45

2. If $r < n$, consider each of the following statements. Is the statement true?

yes	no	**a)**	If r is 20, then n is greater than 20.
yes	no	**b)**	If r is 53, then n is less than 53.
yes	no	**c)**	r could be 14 when n is 16.
yes	no	**d)**	r could be 40 when n is 10.
yes	no	**e)**	r could be 8 more than n.

Questions 3 and 4 refer to the following situation:

The number of times that Christine mows the lawn in the summer is greater than the number of times she shovels snow in the winter. If m stands for the number of times Christine mows the lawn and s stands for the number of times she shovels snow, then in symbols, $m > s$.

3. For each of the given values of m or s in the table below, give two possible values for the other variable.

m		s	
10		_____	_____
12		_____	_____
_____	_____	15	
_____	_____	6	

4. Christine earns $10.00 each time she mows the lawn and $15.00 each time she shovels snow. If she mowed the lawn 9 times in one summer, what is the most she could have earned shoveling snow that winter?

Concept of Equality and Inequality
Practice Sheet 1

Name: _____ A L G E|B|R|I|D|G|E|™

Complete each table with five number pairs that will make the corresponding equation true.

1.
$$36 = 2 \times \square + \triangle$$

Number That Could Go in the $\square$	Number That Must Go in the $\triangle$
_____	_____
_____	_____
_____	_____
_____	_____
_____	_____

2.
$$18 = 3 \times \square - \triangle$$

Number That Could Go in the $\square$	Number That Must Go in the $\triangle$
_____	_____
_____	_____
_____	_____
_____	_____
_____	_____

Concept of Equality and Inequality

Practice Sheet 2

Name: _____

Consider each of the following. Is the equation true?

    yes   no   **1.**   14 forks + 3 spoons + 5 knives + 8 forks + 1 spoon
                         = 22 forks + 4 spoons + 5 knives

    yes   no   **2.**   28 = 16 pencils + 12 pens

    yes   no   **3.**   3 baseballs + 11 footballs + 5 baseballs + 20 footballs
                         = 14 baseballs + 25 footballs

    yes   no   **4.**   8 hamburgers + 5 cheeseburgers + 9 french fries = 32

    yes   no   **5.**   4 cups + 2 mugs + 7 glasses + 3 mugs + 8 cups
                         = 12 cups + 5 mugs + 7 glasses

    yes   no   **6.**   25 math books + 26 history books + 23 math books
                         = 25 math books + 49 history books

    yes   no   **7.**   6 jeans + 5 shirts = 2 shirts + 2 jeans + 3 shirts + 4 jeans

    yes   no   **8.**   13 boys + 16 girls = 4 girls + 5 girls + 7 girls + 5 boys
                         + 7 boys

    yes   no   **9.**   6 records + 14 compact discs + 5 tapes
                         = 4 compact discs + 4 records + 10 compact discs
                         + 2 records + 3 tapes + 2 tapes

    yes   no   **10.**   2 dances + 4 movies + 3 dances + 5 movies
                         = 6 dances + 9 movies

    yes   no   **11.**   3 brothers + 2 sisters + 2 sisters = 4 brothers + 3 sisters

    yes   no   **12.**   18 cars + 6 buses + 4 cars = 28 cars + buses

Concept of Equality and Inequality

Practice Sheet 3

Name: _____

The numbers in the set below are possible replacements for the □.

$$\{ 1, 2, 3, 4, 5, 6, 7, 8, 9, 10 \}$$

Write the number or numbers between the braces that will make each equation or inequality true. Then graph your answers on the number line provided below.

A. $13 - \square = 6$ The set of replacements is   {   }.

B. $13 - \square < 6$ The set of replacements is   {   }.

C. $13 - \square > 6$ The set of replacements is   {   }.

On the number line below, draw one oval around all numbers that make equation A true and label the oval A. Repeat this procedure for inequality B and inequality C.

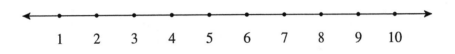

D. $7 \times \square = 28$ The set of replacements is   {   }.

E. $7 \times \square < 28$ The set of replacements is   {   }.

F. $7 \times \square > 28$ The set of replacements is   {   }.

On the number line below, draw one oval around all numbers that make equation D true and label the oval D. Repeat this procedure for inequality E and inequality F.

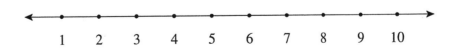

Concept of Equality and Inequality

Practice Sheet 4

Name: _____

In the following table, the set of numbers below each equation or inequality gives the numbers that could be placed in the □ to make each number sentence true.

Complete the table. The first line is done for you.

The replacement set for □ is $\{1, 2, 3, 4, 5, 6, 7, 8, 9, 10\}$.

Equation	> Inequality	< Inequality	Is Entire Replacement Set Used?
1.a) $14 + □ = 17$ $\{3\}$	**b)** $14 + □ > 17$ $\{4, 5, 6, 7, 8, 9, 10\}$	**c)** $14 + □ < 17$ $\{1, 2\}$	**d)** Yes
2.a) $5 \times □ = 35$ $\{7\}$	**b)**	**c)**	**d)**
3.a)	**b)** $10 - □ > 4$ $\{1, 2, 3, 4, 5\}$	**c)**	**d)**
4.a)	**b)**	**c)** $□ + 9 < 14$ $\{1, 2, 3, 4\}$	**d)**
5.a) $2 \times □ + 3 = 9$ $\{3\}$	**b)**	**c)**	**d)**
6.a)	**b)** $□ + 7 - 2 > 10$ $\{6, 7, 8, 9, 10\}$	**c)**	**d)**
7.a)	**b)**	**c)** $30 \div □ < 6$ $\{6, 7, 8, 9, 10\}$	**d)**

Concept of Equality and Inequality

Practice Sheet 5

Name: _____

Answer the following questions.

A jar contains fewer than 18 marbles, and 6 marbles in the jar are red.

1. If □ stands for the number of marbles in the jar that are not red, write an inequality that uses □, the numbers 6 and 18, and two of the symbols $>, <, +, -$ to describe the situation above.

_____

2. What could be the greatest number of marbles in the jar that are not red?

_____

Questions 3 and 4 refer to the following information.

Your favorite rock star has told *Music World Magazine* that he wants to perform fewer than 25 concerts this year. He has already performed 12 concerts this year.

3. If △ stands for the remaining number of concerts that the rock star is willing to perform this year, write an inequality that uses △, the numbers 12 and 25, and two of the symbols $>, <, +, -$ to describe the situation above.

_____

4. What could be the *greatest* number of concerts that your favorite rock star has left to perform this year?

_____

Questions 5 and 6 refer to the following information.

The seating capacity of Nadia's bus is 60 people. After picking up the students at Nadia's bus stop, the bus had 20 boys and 15 girls on it. When the bus arrived at Nadia's school, the total number of girls on the bus was greater than the total number of boys. (No students had gotten off the bus and the seating capacity was not exceeded.)

5. Let △ stand for the number of girls who got on the bus after Nadia's bus stop and □ stand for the number of boys who got on the bus after Nadia's bus stop. Write an inequality that shows that the total number of girls on the bus was greater than the total number of boys on the bus when the bus arrived at school.

_____

6. What could be the *least* number of girls that got on the bus after Nadia's bus stop?

_____

Operations on Equations and Inequalities
Practice Sheet 1

Name: _____   A L G E | B | R | I | D | G | E |™

Expressions 1 to 7 show various ways of combining the terms of the given expression, $8 \cdot \square + 6 \cdot \triangle + 9 \cdot \square + 3 + 7$. Determine which of the expressions 1 to 7 show correct combinations of the terms.

Use the values given for $\square$ and $\triangle$ to evaluate the given expression and the expressions 1 to 7. The first column has been done for you, using 2 for $\square$ and 1 for $\triangle$. For the last column, you or your teacher may choose the values to be used for $\square$ and $\triangle$.

	Value for $\square$	2	1	0	4	
	Value for $\triangle$	1	3	1	2	
Given Expression: $8 \cdot \square + 6 \cdot \triangle + 9 \cdot \square + 3 + 7$		50				
1. $17 \cdot \square + 6 \cdot \triangle + 10$		50				
2. $33 \cdot \square \cdot \triangle$		66				
3. $23 \cdot \square \cdot \triangle + 10$		56				
4. $27 \cdot \square + 6 \cdot \triangle$		60				
5. $17 \cdot \square + 16 \cdot \triangle$		50				
6. $17 \cdot \square + 6 \cdot \triangle + 3 + 7$		50				
7. $20 \cdot \square + 13 \cdot \triangle$		53				

Which of the expressions 1 to 7 always have the same value as the given expression?

Answer: _____

Operations on Equations and Inequalities

Practice Sheet 2

Name: _____

Read and follow the directions given for each question.

1. Decide which of the expressions a to e is another way of writing the given expression, $\square + 4 - 3$, by replacing the $\square$ with the number at the top of each column. Then find the value of each expression a to e. The first column has been done for you, using 5 for $\square$.

Value for $\square$	5	4	6	7
Given Expression: $\square + 4 - 3$	6	5	7	8
a) $\square - 3 + 4$	6			
b) $\square + 3 - 4$	4			
c) $4 + \square - 3$	6			
d) $3 + 4 - \square$	2			
e) $3 - 4 + \square$	4			

Which of the expressions a to e always have the same value as $\square + 4 - 3$?

Answer: _____

2. In each part below write an expression that could represent the events in each situation.

   a) Suppose $\triangle$ represents the number of students in study hall at the beginning of first period. During the first ten minutes, 15 more students arrive because their class has been cancelled. Then, 10 students leave to work on the school newspaper.

   Answer: _____

   b) Suppose $\square$ represents the number of music tapes that Susan has. She gives 5 tapes to her friend Jenna. Then Susan buys 3 new tapes.

   Answer: _____

3. A student was given the following situation:

Suppose □ represents the number of cars on a car dealer's lot. On a certain day, the top salesperson sells 6 new cars and accepts 4 used cars as trade-ins. Write two expressions that could represent the number of cars on the lot at the end of the day.

The student wrote that $6 + \square - 4$ and $4 + \square - 6$ could both represent the situation.

Was the student correct? Explain why or why not.

Answer: _____

_____

_____

Operations on Equations and Inequalities

Practice Sheet 3

Name: _____

For this practice sheet, you will need a 4-function calculator and a scientific calculator. Use the calculators to answer each of the following questions.

1. Enter the expression "$5+2\times3=$" on each of the two types of calculators. Compare the results.

    4-function calculator: _____

    Scientific calculator: _____

    a) Which type of calculator did the addition first? _____

    b) Which type of calculator did the multiplication first? _____

    c) Which type of calculator performs the operations sequentially from left to right?

    _____

    d) Which type of calculator uses the correct order of operations? _____

2. Enter the expression "$6+5\times3+9\times2+6=$" on the scientific calculator. Observe what happens after each entry of a + sign.

    Enter: $6+5\times3+$ _____

    Enter: $9\times2+$ _____

    Enter: $6=$ _____

3. Assuming that the expressions below were entered as written, decide which type of calculator, 4-function or scientific, was used to obtain the value given for each of the expressions. Write 4- function or scientific in the space provided.

    a) $4\times3+5\times2+6=40$ _____

    b) $2+6\times3\times2+4=42$ _____

    c) $10\times2+3\times2+1=47$ _____

Operations on Equations and Inequalities

Practice Sheet 4

Name: _____

1. In each column in the table below, choose a different value for □ . Then evaluate each expression using the chosen value for □ .

Value for □						
Given Expression: $\square \times 6 + 3$						
a) $\square \times 9$						
b) $\square + 3 \times 6$						
c) $\square \times 3 + 6$						
d) $6 \times \square + 3$						
e) $3 + 6 \times \square$						

Which of the expressions a–e always have the same value as $\square \times 6 + 3$?

Answer: _____

2. By 8 p.m. most of the buses of the On-the-Go Bus Company have completed their routes and returned to the parking lot. At 8:15 p.m. one evening, there are 8 buses in each of 3 rows parked in the parking lot. There are 5 more buses that are due in before 9 p.m. Assume that all the buses return as scheduled. Consider each of the following expressions. Could the expression represent the total number of buses that will be parked on the lot by 9 p.m.?

yes	no	**a)**	$8 \times 3 + 5$
yes	no	**b)**	$3 + 5 \times 8$
yes	no	**c)**	$5 + 8 \times 3$
yes	no	**d)**	$3 \times 8 + 5$
yes	no	**e)**	$3 \times 5 + 8$

Operations on Equations and Inequalities

Practice Sheet 5

Name: _____ A L G E |B|R|I|D|G|E ™

Read and follow the directions for each question.

1. Suppose □ stands for a number that makes this equation true:

    $$3 + □ + 7 - 3 = 9$$

    Consider each of the following. Must it also be true for that number?

yes	no	**a)**	$3 + □ + 4 = 9$
yes	no	**b)**	$7 + □ + 3 - 3 = 9$
yes	no	**c)**	$7 + □ = 9$
yes	no	**d)**	$□ + 10 - 3 = 9$
yes	no	**e)**	$□ + 7 + 0 = 9$

2. Suppose $□ - 3 + 7 - 6 + 10 + 2 = 22$ is true when □ represents some number. Answer each of the following questions about the above equation.

    a) What numbers are being added to □? _____

    b) What is the total amount being added to □? _____

    c) What numbers are being subtracted from □? _____

    d) What is the total amount being subtracted from □? _____

    e) Write an equation that is equivalent to $□ - 3 + 7 - 6 + 10 + 2 = 22$ which uses your answers to parts b and d.

    _____

3. Suppose $8 + 2 + □ - 3 + 6 - 4 - 5 = 18$ is true when □ represents some number. A student combines some of the numbers and obtains this equivalent equation:

    $$□ + 10 + 6 - 3 - 4 - 5 = 18.$$

    Write another equivalent equation that could result if the student combined more of the numbers.

    _____

Operations on Equations and Inequalities

Practice Sheet 6

Name: _____

Read and follow the directions for each question.

1. Peter is scheduled to have basketball practice 5 days this week, 5 days the following week, and 6 days the week after that.

   a) Write an expression that could represent the total number of basketball practices Peter is scheduled to have in the next three weeks.

   Answer: _____

   b) Explain why the expression $2 \cdot 5 + 6$ could also represent this situation.

   Answer: _____

   _____

2. Suppose Marsha is scheduled to have basketball practice ☐ days this week, ☐ days the following week, and 4 days the week after that.

   a) Write an expression using addition only that could represent the total number of basketball practices Marsha will have in the next three weeks.

   Answer: _____

   b) Write and expression that uses multiplication and addition that is equivalent to the expression you wrote in part a.

   Answer: _____

   c) Choose a value for ☐ and show that $☐ + ☐ + 4$ has the same value as $2 \cdot ☐ + 4$.

   Answer: _____

3. Match each expression in column A with an equivalent expression in column B.

	Column A	**Column B**
a)	$4 + \square + \square$	$2 \cdot \square - 5$
b)	$\square + 7 + \square$	$4 + 2 + \square$
c)	$2 \cdot \square - \square + 3$	$\square + 3$
d)	$3 \cdot \square - 5 - \square$	$3 - 5$
e)	$\square + 6 + 2 \cdot \square$	$\square^2 + 7$
f)	$7 - \square + 4 \cdot \square$	$3 \cdot \square + 7$
		$2 \cdot \square + 7$
		3
		$3 \cdot \square + 6$
		$3 \cdot \square - 6$
		$4 + 2 \cdot \square$
		$4 + \square^2$
		$8 + \square$

Operations on Equations and Inequalities

Practice Sheet 7

Name: _____   A L G E |B|R|I|D|G|E|™

Answer both parts of each of the following questions.

1. John is on the first step of a ladder and Frank is on the fifth step. Who is higher on the ladder?

    Answer: _____

    Each boy climbs up two more steps. Which boy is higher now?

    Answer: _____

2. In a music store, a record costs $6 and a tape costs $5. Which one costs less?

    Answer: _____

    The store reduces the price of each by $2. Which one costs less now?

    Answer: _____

3. A truck is traveling at 25 m.p.h. while a car is traveling at 30 m.p.h. Which vehicle is traveling faster?

    Answer: _____

    Each vehicle increases its speed by 10 m.p.h. when it gets on the highway. Which vehicle is traveling faster on the highway?

    Answer: _____

4. Elena's puppy weighs 4 pounds while Anita's puppy weighs 5 pounds. Whose puppy weighs more?

    Answer: _____

    Each puppy gains 1/2 pound during the following week. Whose puppy weighs more after that week?

    Answer: _____

5. The temperature at 3 o'clock in Jenkintown was 42° F, and in Yardley it was 38° F. In which town was it colder?

Answer: _____

By 6 o'clock the temperature had fallen 3° in each town. In what town was it colder at 6 o'clock?

Answer: _____

6.

Which point, *A* or *B*, is farther to the right on the number line above?

Answer: _____

Suppose *A* and *B* are each moved one unit to the left on the number line. Which point is now farther to the right?

Answer: _____

7. In the city baseball league, the Bluebirds have won 9 games and the Green Sox have won 7 games. Who has won fewer games?

Answer: _____

The Bluebirds and Green Sox each win exactly 2 more games by the end of the season. Which team ends the season with fewer wins?

Answer: _____

Operations on Equations and Inequalities

Practice Sheet 8

Name: _____   A L G E |B|R|I|D|G|E|™

Rewrite the following equations without parentheses.

1. $2(\square + 1) = 10$ _____

2. $3(4 + \square) = 15$ _____

3. $4(\square - 2) = 12$ _____

4. $(2 \cdot \square + 3)3 = 21$ _____

5. $5(3 \cdot \square - 1) = 10$ _____

6. $2(15 - 4 \cdot \square) = 6$ _____

7. $24 = 4(\square + 1)$ _____

8. $35 = 5(4 - \square)$ _____

9. $3(\square + \square + 2) = 12$ _____

10. $(21 + 3 \cdot \square)2 = 60$ _____